Hervé Chamley

Sedimentology

With 177 Figures and 17 Tables

Translation Thomas Reimer

Springer-Verlag
Berlin Heidelberg New York London
Paris Tokyo Hong Kong Barcelona

Prof. Dr. HERVÉ CHAMLEY
University of Lille I
F-59655 Villeneuve d'Ascq

University of Paris VI
F-75252 Paris cedex 05

ISBN 3-540-52376-6 Springer-Verlag Berlin Heidelberg New York
ISBN 0-387-52376-6 Springer-Verlag New York Berlin Heidelberg

Library of Congress Catalgoging-in-Publication Data
Chamley, Hervé. [Sedimentologie. English] Sedimentology/Hervé Chamley. p. cm. Translation of: Sédimentologie. Includes bibliographical references (p.   ) and index.
ISBN 3-540-52376-6. – ISBN 0-387-52376-6   1. Sedimentology.
I. Title. QE471.C52313 1990 552.5–dc20

This work is subject to copyright. All rights are reserved, whether the whole or part of the material is concerned, specifically the rights of translation, reprinting, re-use of illustrations, recitation, broadcasting, reproduction on microfilms or in other ways, and storage in data banks. Duplication of this publication or parts thereof is only permitted under the provisions of the German Copyright Law of September 9, 1965, in its current version, and a copyright fee must always be paid. Violations fall under the prosecution act of the German Copyright Law.

© Springer-Verlag Berlin Heidelberg 1990
Printed in Germany

The use of general descriptive names, registered names, trademarks, etc. in this publication does not imply, even in the absence of a specific statement, that such names are exempt from the relevant protective laws and regulations and therefore free for general use.

Product Liability: The publisher can give no guarantee for information about drug dosage and application thereof contained in this book. In every individual case the respective user must check its accuracy by consulting other pharmaceutical literature.

Typesetting, Printing and Binding: Brühlsche Universitätsdruckerei, Giessen
2132/3145-543210   Printed on acid-free paper

Preface

Sedimentology, which deals with the geology of sedimentary deposits, has become an actively studied discipline parallel to the development of academic and industrial drilling programs, research on plate tectonic concepts and environmental considerations. Continuous series such as those provided by the Deep Sea Drilling Project or the Ocean Drilling Program constitute outstanding archives to reconstruct environmental conditions governing past seas and adjacent landmasses. Fruitful information arises from investigations performed on thick submarine or exposed series deposited during the successive stages of ocean initiation, deepening and widening (e.g. lithostratigraphy, diagenesis, volcanic and hydrothermal impact). Subduction processes and sea level changes determine renewed interpretations of the sedimentary record on continental margins (e.g. fluid migrations, sequential stratigraphy). Numerous studies involving the dynamics of sedimentary particles are developing in order to better manage the modifications of the shoreline and shelf areas and the commercial use of sedimentary bodies, or to avoid excessive disruptions of natural ecosystems.

"Sedimentology" is especially intended for undergraduate students. The handbook constitutes an introduction to the basic concepts of sedimentation. Chapters 1 to 4 are devoted to the description of the main factors governing the sedimentary processes, and follow the major steps of the surficial geodynamic cycle: terrigenous, biogenic and chemical origins of sedimentary components, characteristics of sedimentary particles, mechanical processes of sedimentation, diagenetic modifications. Chapters 5 to 7 describe the sedimentary environments encountered from high-altitude continental domains to deep-sea areas: glaciers, deserts, lakes, alluvial fans and rivers; deltas and estuaries, coastal and shelf environments; marine environments where various deposits accumulate, such as biogenic oozes, deep-sea clays, nodule-rich and metal-bearing oozes, volcanic and hydrothermal materials, aeolian or glacio-marine sediments. The different environments are described as much as possible with the aid of concrete examples. References are systematically made to present-day situations, and short comparisons are proposed with past geological environments.

The original French edition of this book benefited from the aid of many colleagues and friends. Michel Hoffert, Christian Beck, Jean-Claude Faugères, Pierre Debrabant, Jacques Perriaux, Jean-Joseph Blanc and Alain Desprairies, respectively, corrected and improved several chapters. The entire text was thoroughly read by Pierre Debrabant and Jean-François Raoult. Technical support was provided by Martyne Bocquet, Jean Carpentier and François Dujardin. This book is dedicated to Professors Jean Aubouin, Jean Dercourt and Georges Millot.

Villeneuve d'Ascq, July 1990 HERVÉ CHAMLEY

Contents

1 Origin of Sedimentary Components 1

1.1 Terrigenous Clastic Particles 1
   1.1.1 Introduction 1
   1.1.2 Physical Weathering 1
   1.1.3 Chemical Weathering 2
   1.1.4 General Distribution of Mineral Associations
       Resulting from Weathering 5
   1.1.5 Paleogeographic Application 9

1.2 Formation of Sedimentary Carbonates 12
   1.2.1 Introduction 12
   1.2.2 Main Minerals and Conditions of Formation . . 14
   1.2.3 Primary Chemical Precipitation of Carbonate . . 15
   1.2.4 Organic Contribution to
       Carbonate Sedimentation 16
   1.2.5 Other Factors Contributing to
       Carbonate Sedimentation 19
   1.2.6 General Appearance of Marine
       Carbonate Sediments 23

1.3 Origin of Other Main Sedimentary Components . 30
   1.3.1 Silica 30
   1.3.2 Phosphates 35
   1.3.3 Organic Matter 38
   1.3.4 Evaporites 41

2 Properties of Sedimentary Particles 47

2.1 Grain Size 47
   2.1.1 Scales 47
   2.1.2 Graphic Presentation of Grain Size Data 49
   2.1.3 Characterization of Sedimentary Environments . 51
   2.1.4 Influence of Transport Processes 53

2.2		Shape, Surface, and Orientation of Grains	53
	2.2.1	Shape	53
	2.2.2	Surface Morphology	54
	2.2.3	Spatial Arrangements	55
2.3		Classification of Sediments	57
3		**Deposition of Sediments**	**61**
3.1		Principal Transport Mechanisms	61
	3.1.1	Sedimentary Particles	61
	3.1.2	Flows of Sedimentary Particles	64
3.2		Formation of Basic Sedimentary Structures	66
	3.2.1	Introduction	66
	3.2.2	Formation of Ripples, Current Structures	66
3.3		Formation of Post-Depositional Sedimentary Structures	76
	3.3.1	Erosion Structures	76
	3.3.2	Surface Structures or Imprints	78
	3.3.3	Synsedimentary Deformation	82
3.4		General Mechanisms of Sedimentation and Resulting Deposits	86
	3.4.1	Settling	86
	3.4.2	Gravitational Sliding	88
	3.4.3	Deposition by Density or Gravity Currents	90
	3.4.4	Deposition by Bottom Currents	93
	3.4.5	Deposition by Violent or Exceptional Currents	96
3.5		Identification of Depositional Environments of Ancient Sediments: Potential and Limitations	99
	3.5.1	Introduction	99
	3.5.2	Distance from Coastline	100
	3.5.3	Depth of Deposition	101
	3.5.4	Paleocurrents	102
	3.5.5	Polarity of Beds	103
4		**From Sediment to Sedimentary Rock**	**105**
4.1		Introduction	105

4.2		Formation of Sandstones	107
	4.2.1	Main Stages and Classification	107
	4.2.2	Surface Phenomena	108
	4.2.3	Phenomena of Deeper Levels	109
4.3		Evolution of Clays	111
	4.3.1	Early Diagenesis	111
	4.3.2	Late Diagenesis	113
4.4		Carbonate Diagenesis	116
	4.4.1	Introduction	116
	4.4.2	Calcitic Cement	117
	4.4.3	Conditions for Dolomitization	121
4.5		Evolution of Siliceous Deposits	122
4.6		Formation of Fossil Fuels	124
	4.6.1	Coal	124
	4.6.2	Hydrocarbons	127
5		**Continental Sedimentation**	131
5.1		Glacial Environments	131
	5.1.1	Introduction	131
	5.1.2	Main Types of Glacial Deposits	132
	5.1.3	Glacial Sequences and Environments of the Past	137
5.2		Deserts	139
	5.2.1	Introduction	139
	5.2.2	Recent Desert Sediments	140
	5.2.3	Ancient Desert Sediments	144
5.3		Lakes	144
	5.3.1	Introduction	144
	5.3.2	Recent Lacustrine Deposits	147
	5.3.3	Ancient Lacustrine Deposits	150
5.4		Alluvial Fans	153
	5.4.1	Introduction	153
	5.4.2	Alluvial Fans in Humid Regions	154
	5.4.3	Alluvial Fans in Arid Regions	156
5.5		Rivers and Streams	157
	5.5.1	Introduction	157
	5.5.2	Sedimentary Sequences in Recent Fluvial Deposits	159
	5.5.3	Ancient Fluvial Environments	162

6   Marine Coastal Environments 167

6.1   Deltas and Estuaries 167
     6.1.1 Hydrodynamic and Sedimentational
            Mechanisms 167
     6.1.2 Formation of Delta Complexes 176
     6.1.3 Fossil Estuaries and Deltas 182

6.2   Littoral Environments 184
     6.2.1 Hydrosedimentary Mechanisms 184
     6.2.2 Detrital Environments 189
     6.2.3 Carbonate and Evaporite Environments 197

6.3   Shelf Environments 204
     6.3.1 General Hydrosedimentary Mechanisms 204
     6.3.2 Detrital Environments 209
     6.3.3 Carbonate Environments 215

7   Open-Marine Environments 227

7.1   Submarine Relief and Sedimentation 227
     7.1.1 Introduction 227
     7.1.2 The Continental Margins 227
     7.1.3 Abyssal Plains 229

7.2   Dynamics of Deep-Water Sedimentation 231
     7.2.1 Continental Margins 231
     7.2.2 Open-Marine Basins 240

7.3   Main Deep-Sea Sediments 244
     7.3.1 Carbonate Oozes 244
     7.3.2 Siliceous Oozes 247
     7.3.3 Clays and Argillaceous Oozes 249
     7.3.4 Polymetallic or Manganese Nodules 253
     7.3.5 Metalliferous and Hydrothermal Sediments . . . 256
     7.3.6 Other Sediments 259
     7.3.7 Global Tectonics vs Deep-Sea Sedimentation . . 263

8   References 271

9   Subject Index 281

1 Origin of Sedimentary Components

1.1 Terrigenous Clastic Particles

1.1.1 Introduction

Sedimentary particles supplied from exposed areas are of a diverse nature, but essentially dominated by silicates, which in the coarser fractions are mainly quartz, and in the finer ones mainly clay minerals. These are predominantly the siliciclastic particles resulting from the physical and chemical attack on continental surface rocks by the action of water, temperature, biologic activity, etc. Weathering constitutes a combination of mechanisms liberating rock particles and removing the dissolved elements from the surface of the earth prior to the process of erosion taking over, which leads to transport and eventually to sedimentation. The susceptibility to weathering depends mainly on the type of source rock. Thus a porous sandstone is much more subject to weathering than a recrystallized one, a pure chalk more than a siliceous one, and a surface clay more than a shale resulting from deep diagenesis. A heterogeneous finely laminated or fractured rock will weather more readily than a homogeneous or compact one.

1.1.2 Physical Weathering

In regions characterized by great fluctuations in temperature and humidity, mechanical fragmentation of rocks is the predominant process. It operates mainly through the crystallization of water, or the salts dissolved in it, within fissures or pores of the rock. The *alternation between freezing and thawing* leads to the fragmentation of the rock due to the formation of ice. The acicular ice crystals grow preferentially at right angles to the walls of the fissures by capillary migration of water films to their extremity. This leads to intense pressure in preferred directions, supported by a volume increase of about 10% when the water freezes. The *crystallization of salts* like halite, introduced by marine solutions, or gypsum resulting from the oxidation of pyrite, is also accompanied by expansion, leading to fragmentation, especially of porous rocks. Temperature changes between high and moderate levels, like those encountered in the hot deserts, appear to be insufficient as the sole cause for the fragmentation of rocks which more likely is the result of long-term chemical effects. The phenomenon of *alternate wetting and drying of clayey rocks* from lagoons, marshes, and estuaries is well known. Through preferential evaporation of water along polygonal boundaries of mud-cracks, dry clay flakes are made available to erosion. And last, but definitely not least, *biogenic structures* resulting from the activities of burrowing animals, growth of roots, lichens, mosses, and bacteria add to the fragmentation of outcropping rock formations.

1.1.3 Chemical Weathering

Chemical weathering is much more widespread on the surface of the earth than physical weathering which precedes, accompanies, or outlasts it. It is determined by the necessity of the outcropping rocks to establish an equilibrium with the prevailing surface conditions. This geochemical reorganization results in the formation of new materials at the interface of the lithosphere with atmosphere, hydrosphere, and biosphere. This complex of *weathering reactions* results in a *soil* (Fig. 1.1), made up of a number of levels with specific minerals approaching near-surface equilibrium. This evolution of minerals may be schematically expressed by the following relation:

PRIMARY MINERAL   +   ATTACK SOLUTION
(source rock)   (water of differing salinity)

→   SECONDARY MINERAL   +   RESULTING SOLUTION
    (weathering complex, soil)   (drainage solution)

There are four main processes of weathering, depending on the nature of the attack solution (Table 1.1). Hydrolysis is the most widely developed and best known process. It represents the attack on rocks by pure or slightly CO_2-charged waters at intermediate pH (5 to 9), resulting in a progressive removal of ions from the source minerals. The individual steps of hydrolysis, as in the other weathering mechanisms, depend on the nature of the source rock, the composition of the reagents present in the attack solution, and on climate and topographic relief. In general, hydrolysis is supported by the following factors:

1. preponderance of soluble minerals: the order of destruction of the main silicates follows essentially the order of crystallization of the same minerals during the formation of igneous rocks:

LESS RESISTANT   MORE RESISTANT
Olivine-pyroxene-amphibole-biotite-K-feldspars-muscovite-quartz
Ca plagioclase – Na plagioclase

2. abundance of fine-grained minerals: high specific surface, large number of points of attack;

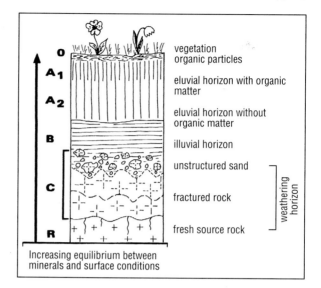

Fig. 1.1. Horizons in a soil profile

Table 1.1. Main chemical weathering processes at the Earth's surface, according to the characteristics of attack solutions (after Pédro 1979)

	$pH < 5$	$5 < pH < 9.6$	$pH > 9.6$
Solutions low in "saline" elements (about N/1000)	Acidolysis waters enriched in soluble organic acids	Hydrolysis waters pure or charged with CO_2	
Solutions rich in "saline" elements (Na, K, Ca …)		Salinolysis waters enriched in salts of strong acids (chlorides, sulfates)	Alcalinolysis waters enriched in salts of weak acids (carbonates, bicarbonates)

3. bacterial activity: supporting the formation of organic acids which participate in the attack on the original minerals;

4. higher temperatures and humidities: acceleration of the weathering processes;

5. good drainage (well-developed slopes): removal of the ions dissolved from the minerals, preservation of undersaturated conditions.

The products of chemical weathering, and notably of hydrolysis, are highly varied. The most common minerals which characterize the majority of soils and also sustain sedimentation, are the clay minerals, accompanied by oxides of iron, aluminum, etc. The *clay minerals* consist essentially of silica, aluminum, oxygen, and the hydroxyl ion OH^-. These compounds are organized in octahedral layers dominated by Al^{3+} (= dioctahedral layers in which two atoms suffice for occupying the six corners of the octahedron) or

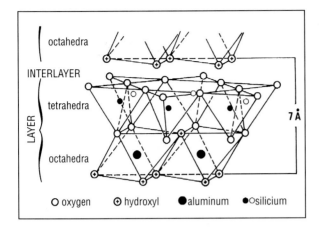

Fig. 1.2. Sheet structure of kaolinite

Mg^{2+} (= trioctahedral layers), and by tetrahedral layers dominated by Si^{4+} and O^{2-}. These *layers* are connected to each other through jointly "utilized" layers of O and OH. This results in a *layered structure* (micaceous structure, phyllosilicates), separated by interlayer spaces. Three main types of structures may be observed (Fig. 1.2): One tetrahedral combined with one octahedral layer (type 1:1 or T.O.: kaolinite family), one octahedral combined with two tetrahedral layers (type 2:1 or T.O.T.: pyrophyllite, illite, smectite, or vermiculite families), or finally a layered structure of the 2:1 type with another octahedral layer in the interlayered position (2:1:1 type; chlorite family). A clay mineral thus consists of a number of stacked layers. The basic structure is frequently complicated by the partial substitution of Si^{4+} by Al^{3+} in the tetrahedra, or of Al^{3+} by Mg^{2+}, Fe^{2+}, and Fe^{3+}, etc. in the octahedra. This can lead to a deficit of positive charges in the individual layers which has to be compensated by the presence of hydroxyl ions in the interlayered positions. The classification of the 2:1 clays is based largely on the degree of this deficit in charges (Table 1.2). In addition, there are clay minerals with X-ray diffraction patterns intermediate between those of two or three simple minerals: the so-called mixed-layer minerals with regular or irregular patterns of interstratification, marking stages of evolution between two different minerals. Furthermore, there are the so-called pseudo-phyllosilicates characterized by continuous tetrahedral layers in combination with discontinuous octahedral layers (with Mg), i.e., the fibrous clays palygorskite and sepiolite (cf. Brindley and Brown 1980; Caillère et al. 1982).

Depending on its intensity, chemical weathering through hydrolysis results in degraded clay minerals (e.g., illite), minerals derived from the transformation of other species (e.g., degraded aluminous smectites), or authigenic minerals formed by ions derived from the weathering solutions (e.g., kaolinites, authigenic Al-Fe smectites). From the increasing intensity of weathering of K-feldspars, three stages of mineral genesis may be identified (Table 1.3) which are marked by an increase of silica solution and the gradual

Table 1.2. Main clay mineral groups

1:1 clays: kaolinite (Al), serpentines (Mg), halloysite (H_2O)
2:1 clays:
- without substitution: pyrophyllite (Al), talc (Mg)
- with tetrahedral substitution by x:
  - $x > 0.6$; interfolia compensation by:
    hydrophobic cations (K, Na): illites, micas, glauconites
    hydrophilic cations (Ca, Mg): vermiculites
  - $x < 0.6$; with possibility of octahedral substitution: smectites
    (montmorillonite and beidellite: Al; nontronite: Fe;
    saponite and stevensite: Mg)

2:1:1 clays: chlorites (mostly Mg), berthierine (Fe)
Intermediate clays: regular or random mixed-layers e.g.: illite-smectite, chlorite-vermiculite, kaolinite-smectite)
Pseudo-sheets (fibrous clays, Mg): palygorskite, sepiolite

Table 1.3. Increasing hydrolysis stages of K-feldspars (from top to bottom)

Primary mineral	Attacking solution	Secondary mineral	Remaining solution
Orthoclase $2.3 \times KAlSi_3O_8$	8.4 H_2O	Beidellite (Al-smectite) $(Si_{3.7}Al_{0.3})O_{10}Al_2(OH)_2K_{0.3}$	$3.2Si(OH)_4 + 2(K^+, OH^-)$
Orthoclase $2 \times KAlSi_3O_8$	11 H_2O	Kaolinite $Si_2O_5Al_2(OH)_4$	$4Si(OH)_4 + 2(K^+, OH^-)$
Orthoclase $KAlSi_3O_8$	16 H_2O	Gibbsite $Al(OH)_3$	(K^+, OH^-)

formation of minerals with two, one, and eventually without any, tetrahedral layers. It is possible to consider the main mechanisms of mineral formation in soils as a function of the intensity by hydrolysis (Table 1.4).

1.1.4 General Distribution of Mineral Associations Resulting from Weathering

The geographic distribution of minerals formed in recent soils and weathering horizons and supplied to sediments by erosion follows certain rules controlled by climatic, morphologic, and geodynamic factors. The recognition of these rules permits the reconstruction of ancient environments in those frequent cases where the particles removed from the exposed lands preserve the essential characteristics of their continental derivation during sedimentation and subsequent burial. For rather coarse particles, from gravel to silt, this information covers petrographic source and geodynamic situation. For the fine particles, the information furnished is more varied and complex, and thus merits a brief review.

1.1.4.1 Parallel Effects of Temperature and Moisture in Well-Drained Areas

Where temperatures are very low and water consequently mostly immobilized in the form of ice, as in *glacial regions of high latitudes or altitudes,* physical weathering will predominate. The minerals supplied to the sediments are those of the source rocks. One finds mostly illite, chlorite, and various non-clayey particles derived from the fragmentation of old continental masses of the northern hemisphere and Antarctica (Millot 1970; Griffin et al. 1968). Locally, in the sediments of the Arctic Ocean, minerals such as Mesozoic kaolinites inherited from ancient sedimentary cover rocks may be observed.

Where temperatures and precipitation are moderate, as in the majority of the *temperate climatic regions,* hydrolysis is sufficiently intense to lead to a degradation of the minerals and to their partial transformation. In the various types of the resulting brown soils there occur illites, irregular mixed-layer minerals (more smectitic than vermiculitic as the humidity increases), vermiculites, and degraded poorly crystallized smectites (pseudo-bisialitization). The formation of kaolinite or gibbsite only affects the more labile minerals such as feldspars, if the removal of the weathering solutions is assured.

In the *warm and humid climates of the tropical and equatorial regions* intense hydrolysis leads to the formation of a thick weathering crust which tends to cause a smoothing of the topographic relief (Millot 1980). The resulting intertropical red soils, or true laterites, are characterized by the accumulation of a ferruginous crust. The authigenic minerals, dominated by kaolinite in association with more or less hydrated metal oxides (goethite, gibbsite) form in abundance and almost independently from the composition of the source rocks. The predominant formation of kaolin-group minerals is referred to as monosialitization, whereas weathering producing aluminum hydroxides and/or oxyhydroxides is called alitization (cf. Table 1.4).

There is thus a latitudinal distribution in the genesis of clay minerals and their associations in weathering horizons and soils, just as temperature and humidity increase jointly from the polar regions to the equatorial zones (Millot 1970). Continental erosion leads to the transport of these minerals to the

Table 1.4. Chemical characterization of hydrolysis steps under surface conditions (after Pédro 1979)

	Partial hydrolysis		Complete hydrolysis
Budget	Incomplete desilication Incomplete de-alkalination	Complete de-alkalination	Complete desilication Complete desilication
Minerals formed	2:1 clays e.g. beidellite	1:1 clays e.g. kaolinite	Al-hydroxides e.g. gibbsite
Interfoliar cations	Na, Ca, K	–	–
Process	Bisialitization (2 layers of Si-tetrahedra)	Monosialitization (1 layer of Si-tetrahedra)	Alitization (absence of Si)

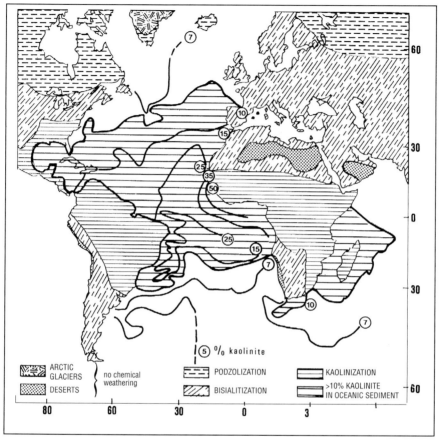

Fig. 1.3. Comparison of kaolinite-rich zones in continental soils (after Pédro 1968) and in marine sediments of the Atlantic Ocean (in %, after Biscaye 1965)

sea where they are deposited as detrital sediments. As a consequence, one may observe that the distribution of the sedimentary marine clays parallels that of the clays in the continental soils from which they have been inherited. Illite and chlorite predominate in the weathering horizons and sediments of the colder regions, moderately degraded minerals in those of temperate regions, and kaolinite in those of the warm and humid zones (Fig. 1.3).

1.1.4.2 Other Climatic Effects

Temperature and humidity sometimes develop in an antagonistic manner. When the climate is cold and humid, as in the *periglacial regions* of high latitude, acid soils rich in organic matter largely complexed with mineral matter are formed on well-drained localities. These are the so-called podzols in which the minerals of the source rocks are intensely degraded by organic acids (acidic weathering or acidolysis), and in which alumina is depleted against silica. Podzolization, marked by the presence of intensely degraded minerals and free silica, is also encountered on certain acid porous rocks in humid temperate and even warm regions.

When the climate is hot and very dry, as in *deserts,* hydrolysis is impeded and the minerals available for erosion result essentially from the mechanical destruction of the source rocks. There is a certain convergence with the weathering processes of high latitudes. The poorly degraded or unaltered minerals inherited by the sediments are those of the rocks cropping out in the corresponding alluvial regions.

When the climate is warm and dry, that is, when it is characterized by a short humid and a long dry season, as in the *warm Mediterranean and dry subtropical regions,* strong hydrolysis alternates with intense evaporation. The cations liberated during chemical weathering organize themselves when the resulting solutions become concentrated, and lead to the formation of well-crystallized authigenic aluminum- and iron-rich smectites (true bisialitization). These minerals in particular are characteristic of the vertisols developed in low-lying, poorly drained zones of arid regions. On more sloped parts of these regions, carbonate crusts (calcretes) are frequently formed (a process referred to as alcalinolysis), a type of contracted soils nourished by evaporation, in which fibrous clays (mainly palygorskite) and more magnesium-rich smectites tend to crystallize (Millot 1980).

1.1.4.3 Non-Climatic Influences

The various climatic effects described above, suggest the suitability of terrigenous clays in ancient sediments for the reconstruction of latitudinal zonation elsewhere on the continents. However, other continental influences may complement or even counteract these effects. This is especially the case for the *weathering of volcanic rocks* in which poorly crystallized compounds rapidly give rise to authigenic smectites in intermediate to high latitudes (Massif Central/France, Iceland). These minerals result from the high weatherability of the rocks and not from a warm climate with contrasting seasons. They may be identified in geological series only by detailed geochemical and electron-microscopic studies.

Another variable concerns the *continental topography and the slope-controlled weathering dynamics.* In warm hydrolyzing climates, kaolinitic soils predominate in higher well-drained regions where the cations, mobilized down-slope, participate in the formation of authigenic smectites in the vertisols. The formation of smectites is supported by areas of shallow relief where the trapped ions are concentrated by evaporation. The authigenic smectites frequently delineate arid climates of lower latitudes, especially in the coastal plains of large discharge basins, such as the Niger. Their formation may then progress in the upstream direction by ionic trapping of the weathering complexes, at the expense of the kaolinitic soils. This leads to the situation that at a certain point, even under the same climatic conditions, one type of weathering may be followed in time by another type (cf. Millot 1980). Under extreme evaporation, low-lying soils may give rise to magnesian clays accompanied by salts as in the Lake Chad basin, a process referred to as salinolysis. When evaporation is limited by intermediate to low temperatures, the soils are swamped and referred to as gleys: hydrolysis is impeded by the lack of removal of the waters and the ions contained in them. Here, the minerals supplied by erosion are only slightly influenced by climatic factors, and correspond closely to those of the source rocks.

In regions subjected to *continuous tectonic instability,* such as the Andes, the process of hydrolysis is continuously counteracted by erosion which removes the weathering horizons as soon as they form. The minerals derived from the soils, and made available to sedimentation, are not in equilibrium with the climate and give the appearance of poorly developed hydrolysis under whatever temperature and humidity. An unstable tectonic situation tends to obliterate the climatic influence on the detrital minerals.

The phenomena of weathering and pedogenesis are of great importance for the formation of mineral associations supplying sedimentary material to deposition. From the situation observed at present, it may be stated that the development of continental soils and their erosion have certainly played an important role in the alimentation of past marine sedimentation, and especially during the non-orogenic periods of earth history. This is illustrated by the abundance of argillaceous formations in the sedimentary series. The presence of ancient climates with stronger hydrolysis than at present may be concluded from the abundance of smectites and kaolinites in pre-Neogene successions and by soils resulting from alitization, such as the Mezozoic bauxites.

1.1.5 Paleogeographic Application

Through their characteristic features, the terrigenous sedimentary particles contribute useful information on the identification of the alluvial sources and the paleogeographic conditions, provided that they do not result from reworking during successive cycles, and that they have not been modified much during diagenesis. From this point of view, the coarser-grained formations are frequently less favorable than the argillaceous ones. Sands and silts have a very long memory of sedimentation which they retain during several phases of reworking, and a very short memory after burial because their permeability favors the circulation of fluids and the attendant geochemical changes.

The optical characteristics of *quartz,* the most common and one of the most resistant minerals in sands, contribute to the identification of its crystalline source rocks. Volcanic rocks contain mostly monocrystalline quartz with straight extinction. Medium- to high-grade metamorphic rocks are richer in monocrystalline quartz with undulatory extinction and polycrystalline quartz consisting of two to three subunits showing identical optical orientation. Low-

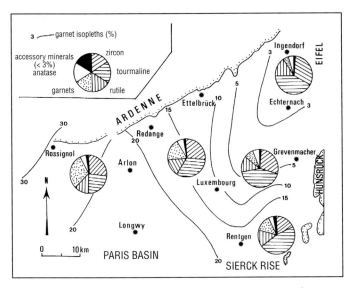

Fig. 1.4. Distribution of main heavy minerals in lower-Liassic Luxembourg Sandstone of the northeastern part of the Paris Basin. Garnets are mainly derived from the Ardenne, zircon and tourmaline essentially from the Eifel (NE) and the Sierck sill (SE). Anatase and rutile are ubiquitous (after Berners 1985)

grade metamorphic rocks contain approximately equal amounts of monocrystalline quartz with undulatory and non-undulatory extinction and of polycrystalline quartz, consisting of two or more subunits. The *feldspars* are less frequent than quartz where chemical weathering is intense, sedimentary reworking frequent, and the depositional energy high. *Heavy minerals* often represent useful indicators of their respective source rocks: tourmalines from plutonic rocks; garnets from schists, the degree of metamorphism of which increases as the Fe- and Mn-contents of the garnets grow at the expense of Ca and Mg; kyanite, sillimanite, and staurolite from micaschists and gneisses. From these indicator minerals interesting paleogeographic reconstructions may be obtained (Fig. 1.4).

When studied in their sedimentological and geochemical context in ancient sequences, the *terrigenous clays* frequently supply information on genetic, climatic, morphologic, and tectonic factors (cf. Millot 1970; Potter et al. 1980; Chamley 1989). This is, for example, the case in the Albian of the North Atlantic, as concluded from the drill holes executed during the Deep Sea Drilling Project (Fig. 1.5). Here, contrasting sedimentary characteristics are separated by a zone situated along the meridian of the present island of Bermuda. To the west, clays derived from flat-lying terrestrial soils are abundant, which were deposited at low rates. Contents of organic matter and residual calcite are low in these sediments. A strong continental influence is indicated by the abundance of aluminum. In contrast to this, to the east, a mixture of clays resulting from different rocks and soils (illite, kaolinite, fibrous clays), was

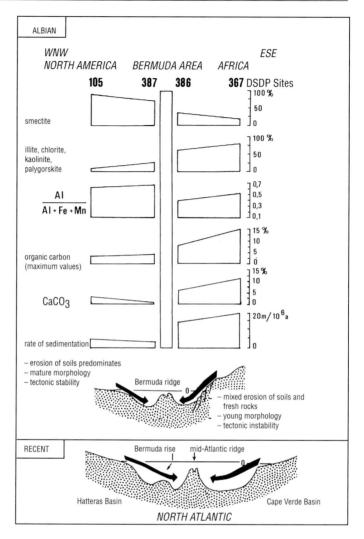

Fig. 1.5. Terrigenous sources and inferred paleogeographic framework of the North Atlantic during the Albian, compared to recent situation (after Chamley and Debrabant 1984). For explanation, see text

deposited at higher rates with high concentrations of organic matter and accompanied by calcareous turbidites. A strong volcanogenic component is indicated by elevated contents of iron and manganese. Symmetric gradients, although at different intensities, are observed for the various parameters when traced at increasing distance from the North American and African coastlines to the open sea. Locally, the submarine morphology of the North Atlantic during the Albian appears to have been marked by the presence of a major barrier, most probably the mid-oceanic ridge, along the Bermuda meridian.

The continental morphology along the borders of the North Atlantic during this period appears to have been rather different: Along the western side, the land was peneplained, permitting the development of poorly drained soils in equilibrium with a moderately warm climate of seasonally fluctuating humidity. As a consequence, weak terrestrial erosion led to slow sedimentation in an oxidizing environment and to rates of carbonate reworking decreasing in the downslope direction. In contrast to this, to the east, the relief was continuously rejuvenated by tectonic instability, leading to active erosion of pedogenic and geologic formations, to rapid marine deposition and an upslope increase in the degree of sedimentary reworking. Rapid burial resulted in a reducing environment, and locally indications of volcanism may be observed.

1.2 Formation of Sedimentary Carbonates

1.2.1 Introduction

Carbonates are mostly of marine origin, resulting from biological and biochemical processes. There are, nevertheless, also various types of freshwater carbonates of biological and biochemical derivation. In recent environments the formation of carbonates results predominantly from planktonic activity, benthic activity being restricted mainly to the shelf areas of intertropical and temperate regions. The development of carbonate platforms during certain notably interorogenic periods such as the Mesozoic, led to the

Table 1.5. Classification of limestones according to the arrangement of constituents (after Dunham 1962; Embry and Klovan 1971)

Structures recognizable							Structures not recognizable
Constituents not connected during deposition						Constituents connected	
Less than 10% allochems > 2 mm				More than 10% allochems >2 mm			
Presence of calcareous ooze (particles < 0.03 mm)		Absence of calcareous ooze		Grains not in contact with each other	Grains in contact with each other		
Grains not in contact with each other							
Less than 10% grains	More than 10% grains	Grains in contact with each other					
Mud-stone	Wacke-stone	Pack-stone	Grain-stone	Floatstone	Rudstone	Boundstone	Crystalline limestone

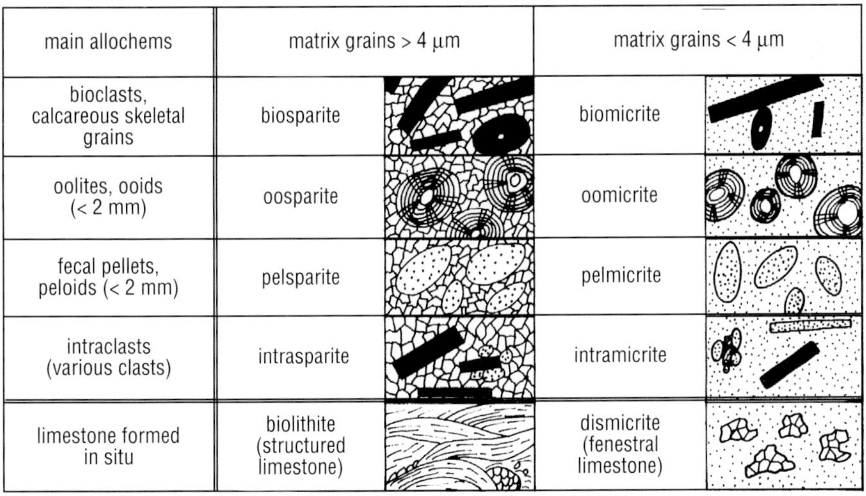

main allochems	matrix grains > 4 μm		matrix grains < 4 μm	
bioclasts, calcareous skeletal grains	biosparite		biomicrite	
oolites, ooids (< 2 mm)	oosparite		oomicrite	
fecal pellets, peloids (< 2 mm)	pelsparite		pelmicrite	
intraclasts (various clasts)	intrasparite		intramicrite	
limestone formed in situ	biolithite (structured limestone)		dismicrite (fenestral limestone)	

Fig. 1.6. Classification of limestones according to their constituents (after Folk 1966)

development of shallow-water carbonates on a much wider scale than at present.

Calcareous oozes are frequently affected by diagenetic reactions at a very early stage, leading to their rapid consolidation. It is due to this susceptibility to diagenesis, that in calcareous rocks distinguishing between primary and secondary features is often possible only with difficulty. This explains the various, essentially descriptive, *classifications* which have been proposed for these rocks, such as those of R.J. Dunham and R.L. Folk (Table 1.5; Fig. 1.6). In most carbonate sediments and the rocks resulting from them, a bimodal grain-size distribution is observed. Usually a distinction is made between individual grains or *allochems* (commonly of a size larger than 63 μm) and the *matrix* of either primary or secondary origin (calcareous ooze vs cement). The distinct calcareous particles consist of skeletons (tests), skeletal debris (bioclasts) and non-skeletal grains such as ooids, fecal pellets, algal encrustations, and aggregates. The detrital particles derived from pre-existing carbonates (lithoclasts) are subdivided into intraclasts derived from neighboring, only slightly older sediments (e.g., beach sands, consolidated micritic oozes), and extraclasts from the erosion of considerably older rocks. The matrix may contain grains still recognizable under the optical microscope as being of biological derivation (e.g., coccoliths) or of chemical origin (sparry cement, consisting of larger interlocked calcite crystals). Also present may be very small calcite crystals (below 4 μm) forming homogeneous domains of micrite.

In the carbonate nomenclature proposed by R.L. Folk, the combination of these terms permits taking into account the presence of the two types of detrital grains (e.g., biopelsparite, bio-oomicrite) and of matrix types of different sizes (e.g., biosparrudite, intramicrudite). The numerous in-depth studies initiated notably by hydrocarbon research resulted in excellent treatises on the subject (e.g., Milliman 1974; Purser 1980, 1983; Scholle et al. 1983).

1.2.2 Main Minerals and Conditions of Formation

In salt water, mineral substances are dissolved about 300 times faster than in freshwater. However, this does not apply to all chemical elements. Calcium, in particular, is relatively less abundant than the other main dissolved elements (Na, K, Mg) in marine waters, because of its strong extraction by organisms.

The two main modifications of calcium carbonate, viz. calcite and aragonite, may form through biochemical as well as inorganic processes. *Aragonite* is not precipitated from freshwater in any great measure and is thus characteristic of marine environments. It is a mineral of low stability and equilibrates only under conditions of high-pressure metamorphism such as in the blue-schist facies. In sediments, it tends to change to calcite, causing the diagenesis of carbonates to be intense and rapid. Because of this, aragonite is rare in ancient sequences. In addition, the extraction of Ca from sea-water increased in the course of geological time (at least up to the Mesozoic), especially through the development of calcareous organisms. The corresponding in-

Table 1.6. Carbonate mineralogy of main calcareous organisms (+ Common mineral; = Occasional mineral)

Organisms	Aragonite	Low-Mg calcite	High-Mg-calcite	Aragonite and calcite
Molluscs				
Cephalopods	+		=	
Gastropods	+			=
Bivalves	+	+		+
Brachiopods		+	=	
Corals				
Scleractinids	+			
Rugosae and Tabulates		+	+	
Sponges	+	+	+	
Bryozoans	+		+	+
Echinoderms			+	
Ostracods		+	+	
Foraminifera				
Benthic	=		+	
Planktonic		+		
Algae s.l.				
Coccolithophoridae		+		
Rhodophyceae	+		+	
Chlorophyceae	+			
Charophytes		+		

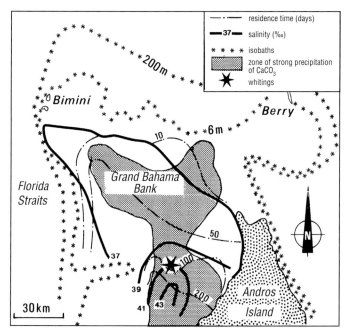

Fig. 1.7. Conditions for precipitation of aragonite in shallow, stationary, and supersaturated waters of the Bahama Bank (after Broecker and Takahashi 1966)

crease of the Mg-Ca ratio actually has lead to the formation of aragonite in preference to calcite in recent marine environments (Berner et al. 1978).

Marine calcite includes Mg-poor (below 4% $MgCO_3$) and Mg-rich terms. The latter usually have contents of 11%–19% $MgCO_3$, but tend to lose some of this with time. Aragonite generally has low Mg-concentrations, but may contain up to 1% strontium, substituting for calcium. The mineralogy of recent calcareous sediments depends on the nature of the organisms concerned (Table 1.6), and on the temperature of the sea-water. Recent shelf carbonates are rich in aragonite and high-Mg calcite in the intertropical regions, becoming richer in low-Mg calcite in the more temperate zones. Slightly ferriferous calcites may form under slightly reducing conditions.

Dolomite, the mixed carbonate of Ca and Mg, appears during diagenesis (cf. Chap. 4). Finally, it should be noted that recent and ancient carbonates are nearly always associated with other minerals such as clay, opal, quartz, pyrite, hematite, and phosphates.

1.2.3 Primary Chemical Precipitation of Carbonate

The direct precipitation of carbonates from ions dissolved in sea-water represents a rare mechanism, known only from certain tropical and subtropical shallow-water environments such as the Bahama Bank, the Persian Gulf, and the Dead Sea. When the surface layers of the sea are supersaturated against the various carbonates, only *aragonite* appears to be prone to precipitation. The process is effective only in waters of elevated salinity and temperatures, and with longer residence times (Fig. 1.7). The precipitation of aragonite in tiny acicular crystals frequently leads to the formation of milky clouds (so-called "whitings") in the water. They appear to originate virtually always when calcite crystals precipitated over the bottom become suspended in the water, except where, as in the Dead Sea, one can observe a rapid decrease in the HCO_3^--content of the water according to the following formula:

$$2 HCO_3^- + Ca^{2+} \rightarrow CaCO_3 + H_2CO_3.$$

The direct precipitation of calcite appears to be inhibited by elevated contents of dissolved Mg. The precipitation of dolomite is equally inhibited, notably by the slow rate of crystal growth controlled by the high degree of lattice ordering. The recent formation of dolomite is largely restricted to a small number of environments such as evaporitic lakes, the Bahamas, and the Persian Gulf. Being mainly tied to the availability of surfaces that are intermittently falling dry, it is frequently poorly ordered (proto-dolomite), richer in Ca than in Mg, and always a replacement of sedimentary compounds such as foraminiferal tests (cf. discussion in Purser 1980; Leeder 1982; and Chap. 4).

1.2.4 Organic Contribution to Carbonate Sedimentation

Numerous carbonate rocks result from the *accumulation of organic skeletal constituents,* the size of which varies from large entities, such as mollusc shells or particles resulting from their breakage (bioclasts), to coccoliths (measuring only a few micrometers) in the sedimentary matrix. Most of the skeletal debris is derived from the hard elements of invertebrates. The associations identifiable in the rocks are closer to the initial biological association when the hard components are numerous, the mineralogy of the tests dominated by calcite, when the intensity of biological, physical, and chemical fragmentation is low, and the diagenetic evolution inhibited. The nature of the various groups of calcitic tests becomes modified with time (Fig. 1.8). Furthermore, the true skeletal associations of a given environment may be altered considerably with geological time. The calcareous fauna and flora of Paleozoic reefs differ greatly from that of modern reefs, although the morphological adaptations of various groups of organisms to this agitated environment are comparable. As a result, one finds a great diversity of petrographic facies in biogenic limestones, necessitating detailed studies in thin sections. There are, however, a number of constructional principles in biogenic carbonates which may be used in the identification of ancient depositional environments.

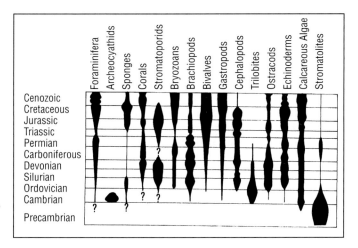

Fig. 1.8. Stratigraphic variation of distribution and taxonomic diversity of main skeletal organisms through geologic time (after Horowitz and Potter 1971)

Bivalves: fresh- and salt-water molluscs participating in the formation of shallow-marine sediments, especially since the Mesozoic, after the decline of the brachiopods. They occur in widely differing environments: within the sediment (endobiontic) or on the surface (epibiontic), in the water in swimming (nectonic) or floating (planktonic) form, or in reef-building colonies (oysters, Cretaceous rudists of the Tethys). Their shells are predominantly aragonitic and are made up of several layers with a characteristic microstructure. During diagenesis they are homogenized to sparry calcite. Molluscs, with calcitic shells such as the oysters, maintain their original structures.

Gastropods: made up of unstable aragonitic shells, appearing mostly as mobile benthos over unconsolidated shallow-marine bottoms. In other environments they are represented by fewer species characterized by a larger number of specimens: over unconsolidated bottoms in fresh, brackish, or hypersaline waters, on hard-grounds (serpulide tubes, borings and reefs built by worms), in marine waters (Cenozoic pteropods of intermediate latitudes).

Cephalopods: common up to the Cretaceous, are exclusively marine and characteristic of the open sea. The shells of ammonites, nautilids, and belemnites consist generally of aragonite now converted to sparry calcite.

Brachiopods: predominantly as sessile benthic, shallow-marine organisms, at present less abundant and occurring in waters of various depths. They consist mostly of a thin external layer of calcitic fibers at right angles to the surface of the shell and a thicker inner layer in which the fibers are orientated at various angles, mostly parallel to the surface.

Corals (cnidaria): present mainly in reef-building (hermatypic) forms in symbiosis with zooxanthellic algae, requiring shallow, warm, and sufficiently clear water. In combination with red algae, their colonies are responsible for the construction of modern reefs (biolithites, boundstones). In reworked form and as isolated corals they support the growth of the reef itself, of the areas surrounding the reef, and occur in various other environments. Less widely distributed are the so-called ahermatypic corals of sometimes rather deep waters (over 1000 m in the North Atlantic), with lower temperatures (10° C and less), where they rarely form colonies. Rugose and tabulate corals with remarkably well-preserved calcitic skelets predominated during the Silurian and Devonian. The scleractinid corals, which consist of unstable aragonite, were widespread during the Triassic. The coral structure, irrespective of the actual mineralogy, is characterized by thin fibers orientated in various directions.

Among the *echinoderms,* only echinoids and crinoids contribute to marine calcareous sedimentation. In recent environments, the crinoids occur mainly in shallow water where they are of little importance. During the Paleozoic and Mesozoic, however, they occupied mostly the carbonate shelves, and the debris derived from them supported bioclastic sedimentation at their growth sites, as well as in the open sea. The echinoderm shells are easily recognized because of their large monocrystalline calcite crystals pierced by numerous pores.

Bryozoans at times played an important role in the sedimentation of reefs and on shelves, notably during the Paleozoic. Consisting originally of calcite or aragonite, their tests appear in highly diverse shapes, most commonly as flattened reticulated aggregates.

The *foraminifera,* protozoans of large size (0.05 to above 10 mm), abound in marine temperate to warm waters where they live partly in planktonic and partly in benthic communities. The former contributed prominently to the sedimentation of "Globigerina" oozes, chalks, and marls in post-Cretaceous oceanic basins of medium depth. Benthic communities are common in the shallow seas of temperate and warm regions where they characterize a number of highly varied sedimentary environments. Consisting essentially of calcite with variable Mg-content, the foraminifera occur in highly diverse structures of agglutinating, microgranular, porcellaneous, or fibro-radial nature.

Other animal species participate in the formation of calcareous sediments only in a more sporadic manner. Calcareous sponges and their relatives, the archaeocyathids, are of importance in certain reef types from the Cambrian to the Triassic. The stromatoporids, a poorly defined group of organisms related to the porifera, built abundant colonies of diverse shapes in Silurian and Devonian limestones together with tabulate and rugose corals. The crustaceans are mainly represented by the trilobites in the Paleozoic, and by the ostracods which are reliable indicators of depositional environments along the continental margins.

Algae have played an important role in carbonate sedimentation ever since the Precambrian. They occur as debris, in sedimentary layers, and as perforating organisms. The red algae (rhodophyceans), such as the corallinacea which are known since the Carboniferous, are made up of a regular cellular structure. Being highly magnesian in composition, they grow preferentially in warm waters under moderate light conditions where, as rhodolites, they frequently encrustate the ocean floors. They play an important role in the formation of cement in modern reefs.

Green algae (chlorophycea) are known since the Paleozoic. Being only moderately calcified, they are easily decomposed and only some of their elements become fossilized. They permit the distinction of freshwater or brackish environments (characea), of tropical lagoonal shallow-water conditions (dasycladacea), or of shallow environments on carbonate platforms (cordiacea: Halimeda, Penicillus). The disintegration of the thin aragonitic needles of the chlorophycea by current action, together with direct chemical precipitation, contributes to the formation of micritic oozes on platforms such as the Bahamas. The planktonic coccolithophorides (chrysophycea) are known since the Jurassic. Their skeletons consist of numerous disc-shaped plates, the coccoliths, made up of radially orientated low-Mg calcite. Being only about 10 μm in diameter, the coccoliths constitute the matrix of the foraminiferal oozes and make up the majority of the calcareous oceanic sediments of various latitudes, such as the Cretaceous chalks and the Tertiary nannofossil oozes.

1.2.5 Other Factors Contributing to Carbonate Sedimentation

Various biological and biophysical processes accompany or favor the direct production of carbonate particles. These mechanisms result in *non-skeletal grains* or in a modification of pre-existing grains.

1.2.5.1 The Role of Bluegreen Algae (Cyanophycea)

Although rarely calcified themselves, the cyanophycea play an important role in the formation of carbonate sediments. They are mostly responsible for the formation of a *dark micritic layer* on the periphery of numerous carbonate fragments. This micritization results from the repeated marginal perforation of these grains by endolithic bluegreen algae and the subsequent filling of this fine network of cavities by calcareous ooze locally enriched in organic matter. It is thus an alteration phenomenon which may continue to the interior of the particle up to the complete micritization, resulting in a *peloid* without recognizable structure. It must be also pointed out that other organisms are able to dissolve limestones. These include bivalves (e.g., Pholas), gastropods, Tunicea, worms (Polychetes), and sponges (e.g., Cliona). However, the effects of these organisms are generally on a larger scale and the results are poorly suited for secondary micritization. The presence of algae in sediments serves as an indicator for the photic zone (100–200 m depth) and thus as a paleobathymetric criterion insofar as this type of micritization can be attributed to bluegreen algae and not to fungi which live in more variable depths.

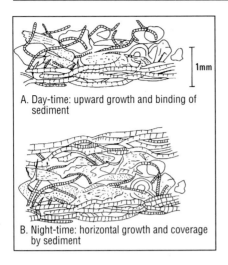

Fig. 1.9. Schematic growth cycle of certain stromatolites (after Gebelein 1969)

A. Day-time: upward growth and binding of sediment

B. Night-time: horizontal growth and coverage by sediment

The cyanophycea participate also in the formation of *algal mats* and *stromatolites*. In intertidal and supratidal zones, in marshes and along lake shores in certain low-latitude regions such as the Bahamas, the Persian Gulf, and Western Australia, the sediment surfaces are covered locally by crusts or domes, mostly consisting of detrital or precipitated material bound together by the action of bluegreen algae (Lyngbya, Schizotrix, etc.) and of bacteria. This leads to calcareous structures morphologically intermediate between those of skeletal and non-skeletal grains (Fig. 1.9). Sometimes the stromatolites attain the shape of irregular nodules with concave and convex outlines, the so-called algal balls or oncolites, consisting of algal laminae growing more or less concentrically around a nucleus (Fig. 1.10). It has to be pointed out that they are frequently difficult to distinguish from the coniatoliths, another type of irregular core-bearing nodules (oncoids) which result from physico-chemical precipitation. The morphological differences between the various algal structures result from differences in the environmental conditions such as depth, wave energy, sequence and amplitude of the tides, supply of detritus, etc. Nodular and domal structures correspond to regions of higher energy, in contrast to planar structures which are more indicative of sheltered zones. Stromatolites are known since the Precambrian, their era of maximum development, culminating in columnar and domal forms such as *Collenia, Cryptozoon*, etc. This abundance of the algal formations and their wide distribution in the subtidal zone during the Precambrian could be largely the result of the absence of bottom-feeding animals such as the gastropods.

The stromatolites are formed in successive *laminar layers*, frequently controlled by a diurnal rhythm. The typical structure consists of a millimeter-size alternation of brown-gray layers of filamentous or unicellular algae growing horizontally or vertically, with fine-grained detrital laminae bound to the upper part of the algal network. The development of the detrital layers is supported by a slackening of the rate of algal growth, the binding action of the vertically growing filaments, and by an increased supply of detritus during storms. Only the upper few millimeters of a stromatolitic body, which may be 50 cm thick, actually contain living algae. Algal mats are frequently fragmented into polygons due to desiccation.

1.2.5.2 Formation of Oolites

Ooids are spherical or subspherical grains with essentially convex surfaces, which distinguish them from oncolites with more irregular outlines. They contain a nucleus of diverse nature (quartz, bioclasts) and a cortex. Ooids include principally the *oolites,* the cortex of which consists of numerous thin concentric layers, or of only one layer in the case of superficial ooids. Their diameter is usually less than 2 mm, above which one talks of pisolites, and they may be simple or composite (Fig. 1.10). To these are also assigned the *spherulites* with a radial cortex, as well as the bahamites (pseudo-oolites) with a micritic unstructured cortex. These two types of ooids are essentially of diagenetic origin.

Recent oolites which are mostly aragonitic, develop in a systematic manner in shallow (1–10 m) marine and lacustrine environments of warm regions submitted to strong hydrodynamic agitation in the presence of much dissolved carbonate. Oolites and pisolites are also formed in certain karst environments. The formation of oolites is thus an essentially physico-chemical phenomenon, viz. the precipitation of calcite on the surface of a particle suspended in a tur-

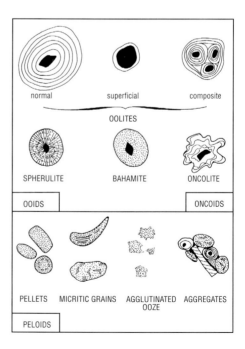

Fig. 1.10. Main non-skeletal limestone grains

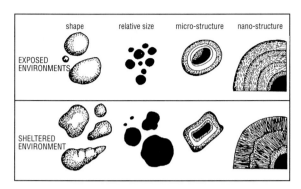

Fig. 1.11. Morphological and genetic relations of oolites (after Purser 1980)

bulent regime. Organic activity is evidenced by the rather frequent development of microvegetation and of insoluble organic residue which participate in the formation of the concentric laminae and are responsible for the diffuse radial structures in ooids, partly by the creation of optimum microenvironments through bacterial activity and partly by calcification induced by amino acids. This influence of organic matter is less pronounced in the areas exposed to wave action than in less agitated environments.

Fossil oolites consist of calcite, and, on the micro-to-nanoscale, are of a radial nature in contrast to the modern oolites in which the fibers grow mainly parallel to the layers of the cortex. This suggests a diagenetic evolution of structure and mineralogy. The question is far from settled. Some recent oolites are calcitic (coastal Baffin Bay, South Texas; northern part of the Great Barrier Reef, Australia; terrestrial caves). However, radial structures presently also form in oolites of little-agitated environments. On average, they are less spherical and show less regular laminations than those of the agitated environments (Fig. 1.11).

1.2.5.3 Excrements and Formation of Aggregates

Homogeneous grains without nuclei or skeletal fragments (*peloids*) consist mainly of fecal particles or *pellets* (Fig. 1.10), in addition to the numerous particles resulting from secondary micritization (Sect. 1.2.5.1). These excrements are abundant in shallow depths of warm and temperate regions. The calcareous particles are produced mainly by gastropods, bivalves, crustaceans, and polychaete worms. They are more clayey in pelagic environments in which, due to settling through the waters down to the sea bed, the preservation of the microorganisms induced by the zooplankton is favored.

The resulting particles are generally small (0.1–0.5 mm), of different shapes according to the type of producers, rich in organic matter, and of dark color. Their identification in fossil rocks is frequently difficult because of the diagenetic modification affecting these little-consolidated particles. Structures comparable to pellets may also result from micritization due to perforation, by agglutination through aragonitic binding, or from fragments of algal mats.

Irregular *aggregates* of peloids, bioclasts and/or ooids, cemented together by aragonite, are found in abundance in neritic sediments such as in the Bahamas. Called *grapestones* as well as *lumps,* these aggregates are formed in little-agitated environments through cementation by sediment-fixing organisms such as slime-secreting algae. Once the intergranular precipitation has started, the aggregate acquires a more regular shape (botryoid lump, "raisin grape"), and may approach a pebble in appearance. There is thus a curious morphological convergence between a calcareous pebble formed in situ, and detrital calcareous pebbles formed by mechanical shaping. It has to be pointed out that mechanical processes may also affect heterogeneous calcareous sediments. This is the case for *mud balls* which result from the collapse of the walls of channels in carbonate environments.

1.2.6 General Appearance of Marine Carbonate Sediments

1.2.6.1 Shallow Water Environments

On the continental shelves, benthic activity is heavily predominant and controls the formation of a considerable variety of carbonate sediments. One possibility of subdividing the neritic calcareous deposits to obtain a reference for fossil deposits consists of the investigation of their dominant constituents, and of the unraveling of their mutual relationships as a function of the environmental conditions. One such tentative subdivision on a global scale has been proposed by Lees (1975), with temperature and salinity as the two main variables. Secondary factors comprise water depth, depth of light penetration, duration of daylight, current conditions, concentration of dissolved matter (notably CO_2) in the water, turbidity, and the nature of the bottom. From these factors there results a latitudinal arrangement of facies of carbonate deposition between 60° N and 60° S as shown in Fig. 1.12.

Skeletal grains: These are dominated by two main associations, viz. by benthic foraminifera and mollusks (*Foramol*) in carbonates of temperate waters, and by bluegreen algae (chlorophyceans) and reef-building (hermatypic) corals (Zoantharia), the *Chlorozoan* association. The Foramol association also comprises cirripedians (crustaceans), bryozoans, and red algae, as well as various other types of organisms of secondary importance. The Chlorozoan association superimposes itself on the Foramol association which, in warmer waters, is always depleted in bryozoans and lacks cirripedians. In places the mollusk component is predominant as on the Paternoster Bank, east of the Indonesian island of Kalimantan (Boichard et al. 1985).

Salinity is of particular importance for the Chlorozoan association which is absent from areas in front of the mouths of large intertropical rivers where salinity is usually below 31‰, but may be developed in semi-closed temperate basins of higher salinity. The preferential development of the Chlorozoan as-

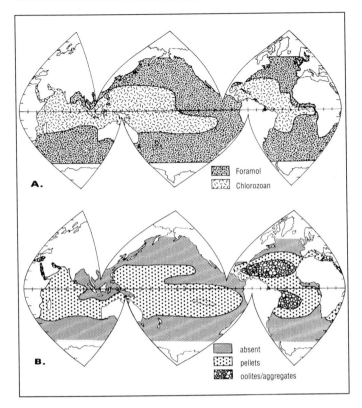

Fig. 1.12 A, B. Main theoretical associations in modern shallow-water carbonates (0–100 m depth). **A** Skeletal grains; **B** Non-skeletal grains. The *lines* cover the oceanic situation irrespective of local peculiarities in certain shallow-water regions such as epicontinental seas and archipelagoes (after Lees 1975)

sociation on the western sides of oceanic basins is controlled by the warm surface currents deflected by the Coriolis force to higher latitudes. An additional association (*Chloralgal*), in which calcareous algae are developed alongside corals, is characteristic of warm waters subjected to fluctuations of salinity like the Bahama Bank.

Non-skeletal grains: These particles without skeletons are absent from areas in which the mean water temperature is below 18° C and the temperature minimum below 15° C, i.e., over a considerable portion of the area of occurrence of the Foramol association. Sediments from the contact zone between the two associations, and also from the Chlorozoan association itself, contain numerous pellets and other peloids. The oolite and aggregate facies are restricted to the tropical regions of the Atlantic and to small basins subjected to strong evaporation leading to high salinities.

The models describing the distribution of recent neritic carbonates, such as the one of Lees (1975), are of great value in the tentative paleogeographic and paleoclimatological reconstruction of ancient environments, notably of the Mesozoic during which such facies attained wide distribution.

1.2.6.2 Deep Water Environments

The calcareous sedimentary material is essentially controlled by planktonic production. This results in rather monotonous sedimentation, dominated by *oozes containing foraminifera, coocoliths,* and sometimes *pteropods*, but subjected to extensive dissolution phenomena. Being little soluble in surface water oversaturated in HCO_3^-, calcite is strongly dissolved in deeper waters, in contrast to the situation for silica (Fig. 1.13). Because of this, the calcareous tests are only slightly altered when they leave the zone of production (mostly the uppermost 500 m) in the direction of the bottom after the death of the original organisms.

The mean water temperature decreases rapidly to 5° C at a depth of about 1000 m, and more slowly below this to values of about 2° C. As the hydrostatic pressure increases concurrently, the relative concentration of dissolved CO_2 also grows, accompanied by a drop in pH. This leads to a *growing dissolution of carbonates in deeper waters* through various successive levels (Fig. 1.14). Down to 3000–4000 m dissolution is slow and regular in the case of calcite, whereas the more fragile aragonite is completely dissolved already at 3000 m.

The *lysocline* of calcite at about 4000 m is the depth below which the rate of dissolution sharply increases. The *critical depth for carbonates* (CCRD = Carbonate Critical Depth) marks the level below which less than 10% of the calcite are preserved. The so-called *carbonate compensation depth* (CCD) is the level at which the complete supply of carbonate is dissolved. Below the CCD there are only siliceous, clayey, zeolitic or metalliferous, oozes depleted in calcium carbonate. The lysocline, defined by the degree of preservation of the mi-

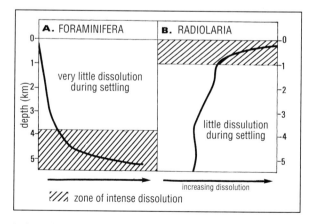

Fig. 1.13 A, B. Profiles of dissolution of (**A**) calcareous and (**B**) siliceous organisms, from experimental data (after Berger et al., in Hsü and Jenkyns 1974)

26 Origin of Sedimentary Components

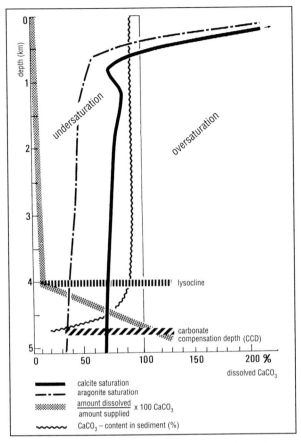

Fig. 1.14. Factors influencing the distribution of $CaCO_3$ in waters of the equatorial Pacific (after Andel et al., in Hsü and Jenkyns 1974)

Table 1.7. Some data related to variations of the Carbonate Compensation Depth

	Position of CCD deeper →		
Factors:			
Productivity			
– Planktonic	– Lower		– Higher
– Benthic (platforms)	– Higher (transgression)		– Lower (regression)
Amount of dissolved CO_2	Higher		Lower
Amount of organic matter	Higher		Lower
Characteristics:			
Mineralogy	Aragonite	Mg-calcite	calcite s.s.
Nature	Small		Large
	Thin		Thick
	Porous		Massive
	Ornamented		

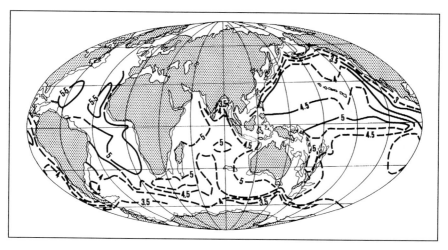

Fig. 1.15. Position of the carbonate compensation depth (CCD) in km (after Berger and Winterer 1974; Burk and Drake 1974)

crofossils, and the CCD, marked by the boundary between carbonate-bearing and carbonate-free sediments, are the most frequently used indicators for paleogeographic and paleoclimatological reconstructions. These two levels vary as a function of a number of factors (Table 1.7).

The environmental factors are supported by mineralogical and morphological characteristics of the tests: their preservation tends to decrease (rise of the CCD) when the planktonic production is diminished (lower supply of calcite), when the benthic productivity increases (reduction of the reservoir of dissolved carbonates), or when the waters become more acidic (due to CO_2 and/or organic matter). As a consequence, the CCD is not at the same level in the various oceanic basins (Fig. 1.15). Especially the oxygen-depleted waters of the Pacific lead to a rise of the CCD to 4200–4500 m, compared to 5000 m in the Atlantic. The marginal oceanic zones are richer in organic matter and thus have a higher capacity for dissolution than the central zones. The identification of the various modifications of the CCD derived from the studies of drill cores of oceanic sediments allow interesting geodynamic reconstructions which are of use for the study of exposed series (e.g., Melguen et al. 1978).

Distribution of oceanic limestones firstly depends on the conditions of dissolution as is clearly evident for the surface sediments of the Atlantic from more than 1700 sampling stations (Fig. 1.16): The deep basins, such as Hatteras and Cape Verde in the North Atlantic, as well as the Brazil, Argentine, Angola, and Cape Basins in the South Atlantic, are considerably depleted in carbonates. Other factors control this distribution. The carbonate content of the sediments increases on the mid-oceanic ridge, its slopes, and on the aseismic ridges of the South Atlantic, zones of relatively shallow depth (2000–3000 m) sheltered from terrigenous supply. It is diminished in front of

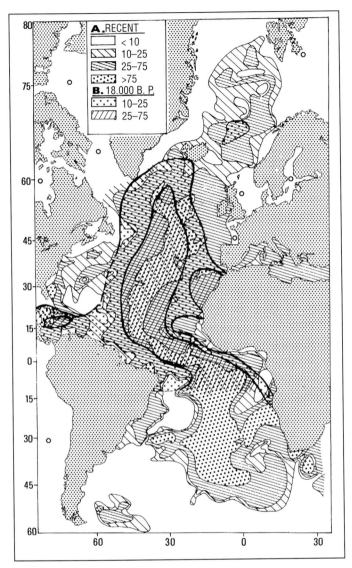

Fig. 1.16. Distribution of $CaCO_3$ (%) in recent sediments (*A*) of the Atlantic compared to situation during the last glacial maximum (*B*, 18 000 B.P.) in the North Atlantic (after Biscaye et al. 1976). o = area not considered

large rivers (Congo, Amazon) as well as in zones of high latitude where carbonate production decreases with planktonic activity, and where there is circulation of cold and relatively aggressive deeper waters (e.g., Antarctic bottom water). This situation was considerably different during the maximum cooling experienced during the last glacial period. At about 18 000 BP the southward migration of the subarctic front led to a decrease of the planktonic

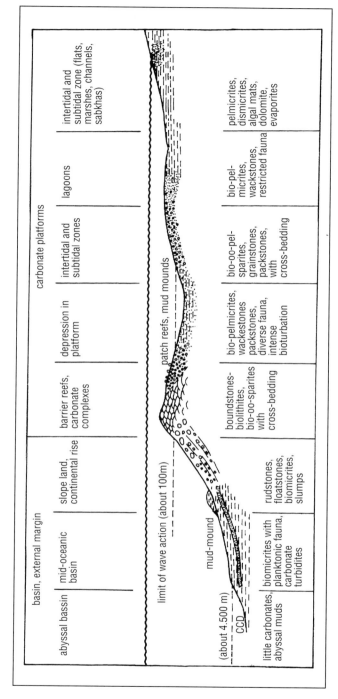

Fig. 1.17. Main environments of formation and depositional facies of marine carbonates (after Tucker 1982)

carbonate production and to increased dissolution of limestone particles on their way to the bottom (Fig. 1.16). In front of the Sahara, the upwelling of cold abyssal waters led to an increase of the carbonate content of the sediments.

1.2.6.3 Zonation

In general, the various carbonate facies are arranged in successive zones from the continents to the open marine bottoms (Fig. 1.17). Paleogeographic reconstructions attempt to establish the characteristics of this zonation during different geological periods. In the supra- and intratidal zones tidal muds, evaporitic deposits, and algal accumulations are the dominant types of sediments. In the infratidal zones skeletal sands as well as bioclastic or chemical muds accumulate. Oolites are always evidence for high-energy environments, participating in the formation of shallow beach bars and occurring also in tidal channels. Reefs and other structures develop on the distal edge of platforms in warm climates. Behind them, fine-grained, bioturbated sediments, frequently rich in organic matter, may accumulate. Isolated reefs or mud mounds, resulting from sediment binding by algae or other organisms such as bryozoans or crinoids, form sporadically on unconsolidated ground behind reef barriers. Products of erosion from reefs and carbonate banks are transported to the neighboring sea bed as a function of particle size and current energy. They frequently supply the slope and foot of the banks with detritus, leading locally to thick mud mounds during certain periods of the past (e.g., Devonian, Carboniferous). In the open sea, calcareous oozes of planktonic origin predominate, at depths not exceeding the carbonate compensation depth.

1.3 Origin of Other Main Sedimentary Components

1.3.1 Silica

1.3.1.1 Origin and Fixation of Silica

Silica in natural waters, derived from siliceous rocks, is supplied essentially from the continents, either directly or indirectly through dissolution of organisms which had used this silica for their own tests (cf. above). Estimates of the quantities involved vary greatly (Table 1.8), but in each case the supply from volcanic activities and submarine hydrothermal systems is much smaller than the supply from surface outcrops where it results mainly from chemical weathering.

The concentration of silica dissolved as H_4SiO_4 in natural waters is usually below 1 ppm in abyssal waters and around 13 ppm in rivers. Under normal conditions of temperature and chemical composition, silica will stay in solu-

Table 1.8. Estimations of the geochemical budget of dissolved silica in the present oceans Data in $10^{14} gSiO_2/yr$ (after Heath 1974; Wollast 1974)

	Supply			Consumption	
	Heath (1974)	Wollast (1974)		Heath (1974)	Wollast (1974)
Rivers	4.3	4.27	Siliceous sediments	10.4	3.60
Submarine dissolution	5.7	0.03	Inorganic precipitation	0.4	0.43
Submarine volcanism	0.9	0.0003			
	10.9	4.30		10.8	4.03

tion as long as its concentration remains below 120 ppm. Consequently, the direct precipitation of silica in natural surface environments is highly exceptional. The bulk of the dissolved silica is fixed by organisms with siliceous tests which represent, as it were, an obligatory stage in the genesis of the majority of the siliceous rocks.

Siliceous sediments of truly chemical origin may be observed in certain lakes of warm climates. Where the pH of the waters exceeds a value of 9 due to intense phytoplanktonic photosynthesis, quartz and certain clay minerals are partly dissolved. The silica thus liberated increases the silica content of the waters which then may be precipitated in the form of a gel consisting of cristobalite when the vegetation dies off (e.g., ephemeral lakes of Australia). In the same way, the evaporation of East African lakes in arid climates leads to waters of high alkalinity rich in silica (up to 2500 ppm) which then is precipitated as magadiite, an unstable hydrated sodium silicate.

The organisms responsible for the extraction of silica from naturally occurring solutions are mainly *diatoms,* especially marine planktonic algae (Jurassic to Present), and *Radiolarians* which are exclusively marine planktonic protozoa (since the Paleozoic). These organisms appear to develop mainly in the surficial plankton as the concentration of dissolved silica is particularly low in marine surface waters. Other silica-fixing organisms are benthic sponges with siliceous spicules and planktonic silico-flagellates. The latter are only of secondary importance.

1.3.1.2 Solution of Silica in Surface Waters

Silica fixed by organisms generally is not very stable. The undersaturation of the surface waters leads to a strong dissolution of the tests (Fig. 1.13) releasing silicic acid:

$$SiO_2 + 2H_2O \rightarrow H_4SiO_4.$$

The majority of the biogenic silica is thus rapidly dissolved in the surface waters, thereby supporting the fresh growth of diatoms and radiolarians. Consequently, a large portion of the silica is involved in a short cycle of biologic fixation and dissolution, characteristic of marine surface waters. The main destiny of the siliceous tests thus is continuous recycling and not fossilization.

As the abyssal waters, due to a lack of silica-extracting organisms, are less undersaturated in dissolved silica than the surface waters, and as the solubility of silica decreases with falling temperature and increasing depth, the conditions necessary for the fossilization of siliceous tests will be encountered only after their transit to water depths of at least 1000 m (Fig. 1.12).

This condition is met where the planktonic production is very high, for example, where the production of siliceous tests exceeds the rate of solution in the surface waters. The siliceous tests arriving at greater depths are rather well preserved because of the absence of a lysocline and a compensation level for silica, and are concentrated against the carbonates which tend to be dissolved at such depths. The siliceous deposits usually are more prominent where the productivity is high and the waters deep.

A comparison of the rates of extraction of dissolved silica by phytoplankton and the distribution of biosiliceous sediments in the oceans of the world clearly illustrates the *close connection between productivity and silica concentration in sediments* (Fig. 1.18). Production is increased wherever upwelling of deep waters rich in nutrients such as nitrates and phosphates takes place: in high latitudes of the northern and southern hemispheres, along the equatorial Pacific, and in front of the west coasts of America and Africa. The concentration of biosiliceous materials in sediments is observed at water depths exceeding the carbonate compensation depth, as in radiolarian oozes of the equatorial Pacific west of the East Pacific Rise, and where calcareous production is restrained by low water temperatures, such as in the diatomaceous oozes of the circum-Antarctic belt, the North Pacific, and the Bering Sea. In contrast to this, planktonic biosiliceous production is partly reduced by extensive biocalcareous production on the East Pacific Rise, and in zones of coastal upwelling, as well as by terrigenous supply in the North Atlantic and parts of the North Pacific.

The application of these findings to ancient successions leads to a wealth of information. It becomes possible to reconstruct the position of ancient zones of productivity and their relative movements resulting from the migration of lithospheric plates through different climatic zones. The radiolarian belts at present are of smaller extent than the diatomaceous belts. However, during the Mesozoic they were much more frequent and covered wider areas. Considering the mainly equatorial distribution of the recent radiolarian oozes, the Mesozoic situation appears to be the result of the wide extent of warmer climates prior to the establishment of a glacial world during the lower-middle Cenozoic.

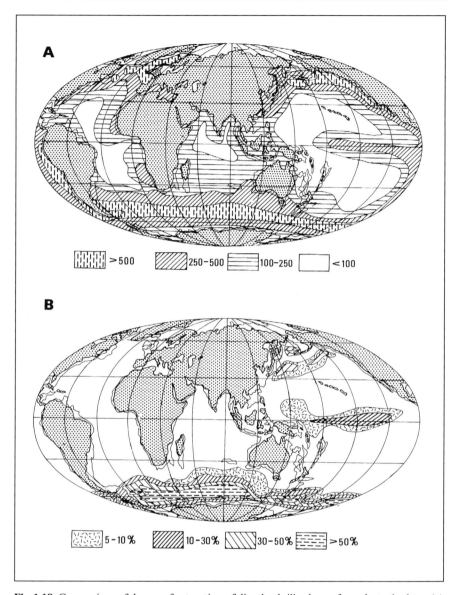

Fig. 1.18. Comparison of degree of extraction of dissolved silica by surface phytoplankton (*A* in 10^{-2} g SiO_2 per year) with content of biogenic silica in surface sediments (*B* corrected to 0% $CaCO_3$) (after Calvert, in Hsü and Jenkyns 1974)

1.3.1.3 Fossilization and Further Evolution of Biosiliceous Tests

The dissolution of siliceous organisms is very active in the surface waters, and does not stop completely in deeper waters (Fig. 1.13). It continues along the water–sediment interface and within the sediment itself. Measurements in the

East Pacific, a zone of high productivity, show that 90%–99% of the biogenic opal are dissolved before reaching the bottom and that on average only about 2% of the original siliceous tests are preserved after burial. As dissolution is highly selective, affecting more the species with thin and ornamented tests, the relation between living and dead associations frequently is much distorted. The following order of increasing resistance to dissolution has been noted: silicoflagellates, diatoms, small- then thick-walled radiolarians, sponge spicules. These effects have to be taken into account for paleoecological and paleobiogeographic reconstructions, and particular attention should be paid to rocks in which siliceous organisms tend to be best preserved: organic sediments and those of low porosity, planktonic fecal pellets not dissociated in the water. It has to be pointed out that abundant biosiliceous sediments, little subjected to dissolution, in places form prolifically in volcanic lakes and in other lakes characterized by water rich in dissolved silica in which diatoms florish (e.g., Massif Central/France).

Biosiliceous particles which escape from the various stages of the surficial silica cycle (Fig. 1.19) are during diagenesis eventually associated with silica resulting from volcanism and particulate matter from the continents (cf. Sect. 1.4.5.1).

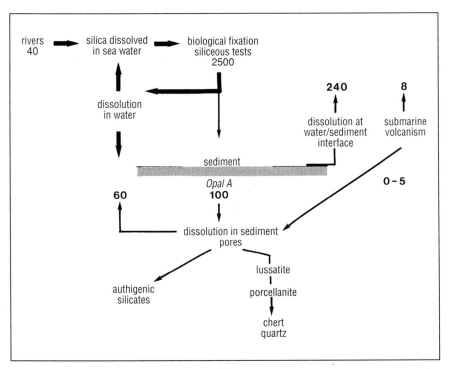

Fig. 1.19. Marine cycle of silica with estimated concentrations of dissolved silica (in 10^{-2} g SiO_2 per year) (after Riech and von Rad 1979)

Radiolarian oozes often change into reddish or jaspilitic radiolarites in which the tests frequently are well preserved. Diagenetic evolution leads further to lydites, blackened by fine-grained organic matter and sulfides, and eventually to graphitic schists (cf. Sect. 4.5). The diatomaceous oozes change into diatomites. The "gaizes" mark the transition between biosiliceous and detritic rocks. The great variety of chemical siliceous sediments, resulting largely from the diagenetic evolution of biosiliceous sediments, should be grouped together under the term "chert".

Cherts comprise the siliceous inclusions in sandstones and especially limestones (flint or "incomplete" flint in marine deposits, "meulière" in siliceous continental limestones). The petrographic evolution from biosiliceous oozes to cherts is accompanied by a mineralogical evolution from disordered opal A to quartz (Fig. 1.19).

1.3.2 Phosphates

Rocks rich in phosphates are collectively referred to as *phosphorites*. The term encompasses rocks particularly rich in calcium phosphates, together with phosphates of iron and aluminum (Slansky 1980). Although phosphates are comparatively sparsely distributed on the surface of the earth (Fig. 1.20), their deposits range in age from the Precambrian to the Cenozoic, and occur in

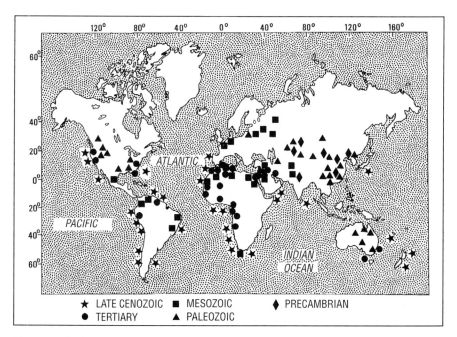

Fig. 1.20. Distribution of main marine phosphorite deposits

small quantities in a large number of rocks characteristically of organic origin. Phosphates are particularly concentrated in vertebrate bones, in durable parts of invertebrates, and in certain natural waste products such as guano and fecal pellets.

The mean concentration of dissolved phosphorus in sea-water is below 0.7 ppm because phosphates are only weakly soluble. The direct precipitation of phosphorus in marine environments thus is hardly imaginable. In warm surface waters its content, resulting from planktonic activity, is 0.003 ppm, whereas in deeper waters it reaches 0.1 ppm. As has been shown for silica, phosphorus is fixed and concentrated essentially by *biologic processes*.

Details of the various mechanisms of phosphatogenesis are still only imperfectly understood, as these minerals ceased forming in notable quantities since the upper Miocene (Slansky 1980; Baturin 1982). Tertiary and Quaternary phosphates are located at relatively low latitudes, i.e., always below 40° latitude, indicating their affinity to *warm waters*. At the same time the paleogeographic study of the large pre-Cenozoic phosphate deposits also shows a preferential development between 0°–40°. Marine phosphorite deposits appear to be connected to the upper part of the continental margins. In recent oceans, this is the zone between shelf and slope. The conditions of formation of these deposits appear to be similar to those of glauconites. In older sequences, they always formed between coastal sediments and those of the open sea. In the Phosphoria Formation of the Permian of the northwestern USA, the phosphorites show their largest development within a 420-m thick sequence of shales, limestones, and dolomites, separating red sandstones and shales, evaporites, and neritic carbonates on the one side from spicule-bearing cherts and deep-water siliceous limestones on the other (Blatt 1982). The upper Cretaceous to Paleogene deposits of North Africa and Anatolia formed on shelves and in peri-oceanic basins intermediate between continental and open-marine environments (Lucas 1979).

The fixation of phosphorus from sea-water is particularly favored in coastal regions along which *upwelling* cold waters supply abundant nutrients from depth. Recent upwellings occur mainly along the west coasts of the continents (western and southwestern Africa, and the Americas) where terrestrial winds (trades, monsoons) drive the surface waters out to the open sea. This leads to a prolific growth of plankton and the biologic extraction of dissolved phosphorus from sea-water.

After the death of the planktonic organisms, their tests, together with the organic matter accumulate massively in the bottom sediments (Fig. 1.21). Trapping and preservation of phosphorus in the sediment require a *slightly reducing environment,* favored by the presence of unoxidized organic matter and only slight submarine relief, somewhat depressed and suitable for accumulation by settling. The coastal basins of Africa present excellent examples for such geochemical traps where in some cases other substances such as hydrocarbons, magnesian clays, silica, etc. may also accumulate (cf. Lucas and Prévôt 1975).

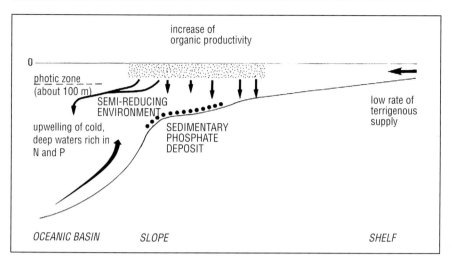

Fig. 1.21. Model of formation of marine phosphorites

The enrichment of phosphorus in sediments is the result of a number of processes:

1. *Exceptional planktonic production,* to which diatoms and dinoflagellates are the main contributors, can lead to mass mortality, the red tide phenomenon, or to increased consumption of plankton by higher organisms, notably fish, and consequently to the increased accumulation of phosphorus on the bottom.

2. In the majority of cases, the absolute concentration of phosphorus appears to be of secondary importance. The phosphorus of planktonic origin may be dissolved in the interstitial pores of the sediments where its concentration then would be much higher than in the overlying sea-water. It could then become fixed as phosphate through gradual oxidation of the associated organic matter. This *precipitation* would occur preferentially at the contact between surface layers rich in oxygen and relatively oxygen-deficient layers (Slansky 1980).

3. A third mechanism, which appears to occur readily under natural conditions, is the *replacement* of calcareous tests by phosphate derived from phosphorus dissolved in the interstitial waters of the sediments (Lucas and Prévôt 1986), a process which may occur in fresh- as well as in marine water. It depends on bacterial activity and is supported by the presence of amino acids.

4. Finally, phosphate particles may be concentrated, either in situ or in reworked form, by the *mechanical action of currents and storms* (epigenetic bioclastic grains, fish bones, etc.)

These mechanisms are active in a more or less successive and complementary manner during the accumulation of phosphates. Their formation requires the enrichment of phosphorus relative to Ca, through planktonic production by mainly chitinous and siliceous organisms, and an enrichment relative to Mg, through early diagenetic formation of dolomite, as often found in phosphorites. The reducing character of the environment appears to be moderate, and intermediate between the two types of rocks frequently associated with phosphates, namely evaporites in which the organic matter is oxidized and the black shales (oozes and dark clayey mudstones) in which it accumulates. Bacterial activity appears to be particularly important in shallow sediments.

To summarize, the formation of sedimentary phosphates requires warm waters, upwelling of oceanic waters rich in nutrients, extensive planktonic production, and the possibility for accumulation of phosphorus in the sediments. These conditions are supported by rather shallow depths which inhibit the dissolution of the organic matter during descent through the waters, and by sufficiently low rates of terrigenous supply in order to avoid rapid burial and dispersion of the phosphate.

Locally, *vertebrates* also give rise to phosphate sediments. The main accumulations, the so-called *bone beds*, result from fish debris like those of the Silurian and Rhaetian of England and the Pliocene of Florida. *Guano* from bird droppings or from chiropteria may locally lead to deposits of economic value, by downward transport in solution and replacement of underlying limestone beds, as observed on some islands of the East Pacific.

The minerals resulting from phosphatization belong mainly to the apatite group, complex calcium fluoro-phosphates of the mean composition (Ca, Na, Mg)$_{10}$ (PO$_4$)$_{6-x}$ (CO$_3$)$_x$ F$_y$ (F, OH)$_2$. The well-crystallized members of this group are francolite (F>4%) and dahllite (F<1%). Collophane is a poorly ordered apatite. Within neutral or slightly acid organic environments, the phosphates are rather stable. Their alteration may lead to particles of various shapes representing the result of a complex mineralogical and geochemical evolution. Their fossilization is accompanied by a relative enrichment of phosphorus as the particles become progressively more resistant.

1.3.3 Organic Matter

Although a ubiquitous constituent of sediments, organic matter is present virtually always only in trace amounts. The mean contents of C$_{org}$ in recent sediments of the North and South Pacific are below 0.25%. In the various basins of the Atlantic the content varies between 0.25%–0.5% and in the Mediterranean between 0.5%–1.0% (Pelet 1985). Contents of only about 1% organic matter are sufficient to give to a sediment a dark-gray to black color. In order to accumulate in notable abundance, and to lead to true organic sediments, the non-mineral constituents of biologic origin have to be supplied by elevated rates of productivity (plankton, coastal terrigenous muds). The bottom waters must be renewed only slowly, or be low in oxygen as in closed or semi-closed basins, or the removal from potential oxidation through sedimentation burial must take place rapidly. Such *conditions are independent of the actual depth*. Organic sediments accumulate in a number of poorly drained continental environments (peats, marshes, lakes) just as well as on the sea bottom, whether close to shore or in the deep ocean (e.g., coastal lagoons, Scandinavian fjords, Gulf of California, Black Sea, Puerto Rico Trench). In the past, organic sediments formed during various geological periods. Of note are the Paleozoic graptolite and goniatite shales, coal, bituminous shales of North America, the "terres noires" of the subalpine Callovo/Oxfordian, oil-bearing Kimmeridgian clays of the North Sea, the Cretaceous black shales of the Atlantic and the western part of the Tethys, and Quaternary sapropels of the eastern Mediterranean. Chemical elements are sometimes concentrated in economic or trace quantities by the organic matter; this is the case with copper and zinc, as well as traces of arsenic, molybdenum, lead, uranium, vanadium, etc., in the Permian "Kupferschiefer" extending through central Europe from Poland to England.

In the sea and in lakes, the massive oxygen depletion in the water, precondition for the preservation of organic matter, may occur in two principal situations (Fig. 1.22).

Stratification of water (A). In basins which are isolated from the open sea, or which have only limited connection with it, the exchange of waters of different depths is impeded. This leads to a stratification of waters in separate layers and to rather stagnant conditions. It is amplified by the fact that specifically lighter river waters flow over the saline waters as in the recent Black Sea or in the eastern Mediterranean during the great pluvial phases of the Quaternary. The oxygenated surficial waters and sediments rapidly grade with depth into anaerobic environments rich in H_2S, conducive to the accumulation of organic matter. A comparable situation occurs when abyssal ocean waters, after travel over large distances, become depleted in oxygen because of benthic activity. This is the case in the Pacific where the dissolved oxygen con-

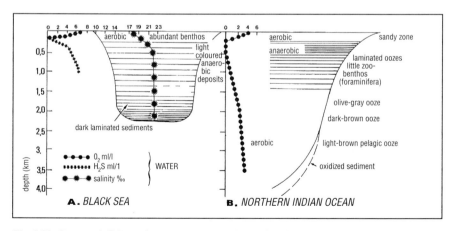

Fig. 1.22. Oxygen deficiency in water masses and associated sediments (after Thiede and van Andel 1977)

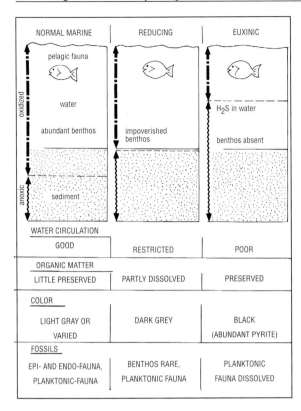

Fig. 1.23. Degrees of stagnation in water masses and resulting sedimentary characteristics (after Tucker 1982)

tent of the Antarctic bottom waters is reduced from 8 ml/l to 2 ml/l during migration to the northern part of the basin (in Kennett 1982). All intermediate stages between open and euxinic environments may be observed (Fig. 1.23). The sediments richest in organic matter are characterized by the absence or at least a great depletion of life on the bottom and within the sediment. The frequency of thin sedimentary laminations is often controlled by the episodic intense development of specific types of plankton (so-called blooms), by the dissolution of calcareous tests and sometimes of the clay minerals by organic acids, leading to the relative enrichment of siliceous tests and plant matter. Laminations can also be marked by the development of scattered sulfides (pyrite, marcasite) in the form of tiny rods, plates, or framboids (cf. Schlanger and Cita 1982).

Oxygen minimum zone (B). In certain basins, high planktonic productivity leads to a strong consumption of oxygen as in the northwestern part of the Indian Ocean. The oxygenation of the intermediate water masses is normally not as efficient as in the surface waters where winds and currents exert a continuous mixing action. As a result, a layer of poorly exchanged oxygen-deficient water is established, leading to the reduction of sediments in contact with it. Still deeper down, the water-sediment interface becomes oxidizing again. The planktonic and benthic activities which would consume oxygen are diminished in the intermediate water because the rate of consumption of oxygen exceeds that of the replenishment. As a consequence, the deeper water deposits may be more oxidized than those of intermediate depths, and reducing sediments may develop above the minimum oxygen layer. These reducing sediments, formed at intermediate depth in oxygen-depleted waters, may be reworked by density or gravity currents and redeposited on the bottom of the basins within normal stratified sedimentary environments. This explanation was proposed by Robertson and Bliefnick (1983) for the majority of the black shales deposited during the Cretaceous in the mid-western Atlantic basin.

1.3.4 Evaporites

Evaporites are essentially sediments of *chemical origin* indicative of *warm climates*. Their three main constituents, halite, anhydrite, and gypsum, may form in marine as well as in non-marine environments. Other, accompanying minerals are frequently indicative of marine or continental conditions (Table 1.9). Lacustrine evaporites of a highly varied nature are known from recent environments, such as certain African lakes and the Great Salt Lake/USA, as well as from fossil sequences like the Eocene Green River Formation/USA. They are, however, always of rather limited extent. The great evaporite occurrences in the Cambrian, Devonian, Permian, Triassic-Jurassic, and Miocene are marine in origin. They are preserved on the present-day continents, e.g., in the Salina Basins of the USA, the Zechstein of northern Europe, in Siberia, and in Anatolia, and below the sea floor in Messinian basins of the Mediterranean, the Miocene of the Red Sea, and the Jurassic of the Gulf of Mexico. Their geodynamic situation is varied. They may have formed in intracratonic basins invaded by the sea like in Permian formations of the European Zechstein, but mostly they formed along con-

Table 1.9. Major saline evaporitic minerals

Marine environment		Continental environment
	Halite NaCl	
	Anhydrite $CaSO_4$	
	Gypsum $CaSO_4 \times 2H_2O$	
Sylvite KCl		Epsomite $MgSO_4 \times 7H_2O$
Carnallite $KMgCl_3 \times 6H_2O$		Trona Na_2CO_3, $NaHCO_3 \times 2H_2O$
Kainite $KMgClSO_4 \times 3H_2O$		Mirabilite $Na_2SO_4 \times 10H_2O$
Polyhalite $K_2MgCa_2(SO_4)_4 \times 2H_2O$		Thenardite Na_2SO_4
Bischofite $MgCl_2 \times 6H_2O$		Gaylussite Na_2CO_3, $CaCO_3 \times 5H_2O$
Kieserite $MgSO_4 \times H_2O$		

Table 1.10. Example of evaporitic sequence, coastal sabkhas of northwestern Africa

Mechanism	Minerals
Dilution by continental run-off	Hydration of gypsum
Shore of lagoon	Contorted enterolithic anhydrite
Intense chemical precipitation	Anhydrite nodules, with enclosures of host sediment ("chicken wire" structure)
Increase of evaporation (salinity 260‰, summer temperature $<35°$ C)	Elongated anhydrite crystals
Evaporation of sea water	Gypsum (rosettes, lanceolate) calcite, protodolomite

tinental margins in the associated subsiding basins, where the evaporitic sequences are intercalated with limestones and various shales. Such formations are tied either to the opening of a basin as in the lower Tertiary of the Alsace/France, the recent East-African lakes and the Dead Sea, or to the marginal areas of the lower Cretaceous Atlantic Ocean, or to the closing stages as in the Mediterranean at the end of the Miocene (Kirkland and Evans 1973; Kendall, in Walker 1984).

Normal marine water is undersaturated with respect to the various evaporitic minerals. In order to facilitate their precipitation, evaporation has to exceed the rate of rainfall and supply of fresh continental or marine water. This situation presently is observed in coastal evaporitic lagoons in warm and dry regions, such as the coastal sabkhas of North Africa (Table 1.10) or in semi-closed gulfs, such as the Kara Bugas of the Caspian Sea. As sea-water is less saturated with sodium chloride than with calcium sulfates and carbonate, the precipitation of halite is slower than that of gypsum or calcite. The salts least concentrated in sea-water are the chlorides and sulfates of potassium and magnesium which consequently will crystallize last. Figure 1.24 illustrates the theoretical sequence and the related conditions for the precipitation of the main salts during evaporation of sea-water. The apparent differences in this sequence as observed in evaporitic series relate mainly to the relative scarcity of potassium salts such as sylvite. Causes for this could be incomplete evaporitic cycles, geochemical disequilibria in the brines, and diagenetic effects.

The evaporation of a given volume of sea-water is insufficient for the formation of a thick evaporitic body. It has been estimated that the evaporation of even the complete mass of the present oceans would only result in a salt layer of about 60-m mean thickness. The evaporation of a 1000-m thick layer of sea-water would only give about 12.9 m of halite. The volume of the salts precipitated from the Permian Zechstein over a restricted area of Northern Europe at a thickness of about 1000 m, represents by itself already about 10% of the total presently available salt reservoir of the oceans, or about 2.4×10^6 km^3. For a basin of limited surface area it is thus imperative to assume the repeated evaporation of waters replenished either from the open ocean or the surrounding continents. This is the case especially for the locally

Origin of Other Main Sedimentary Components 43

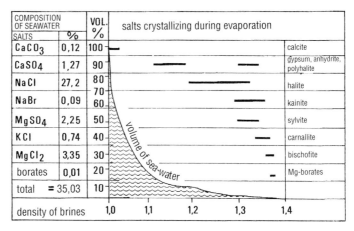

Fig. 1.24. Effects of sea-water evaporation on resulting brines and saline precipitates (after Valyashko, in Richter-Bernberg 1972)

more than 1000-m thick salt deposits of the upper Messinian in the Mediterranean basin which had a restricted connection with the Atlantic Ocean (Hsü et al. 1978).

Recent environments cannot serve as examples for the enormous evaporitic basins of the geological past, as recent evaporites are mostly connected to lagoons or deltas. As a consequence, we are hampered in the formulation of genetic hypotheses. Three types of basins may usually lead to the genesis of thick evaporitic sequences (Fig. 1.25):

The *shallow basins with thin water cover* (A) correspond to giant sabkhas in which subsidence is controlled by the mass of the crystallizing salts. Along their margins supratidal evaporitic lagoons can be developed with deposition prograding in the direction of the open sea. These deposits usually exhibit directional sedimentary structures controlled by currents and wave action. This type of basin does not appear to have been developed much in the past because thick evaporitic deposits require notable subsidence hardly compatible with low topographic relief.

Deep basins with thin water cover (B) are isolated from the open ocean by an elevated ridge which will permit the periodic influx of marine waters and through which water may seep into the basins. Such deposits show sedimentary structures of immersion and resedimentation as the continental influx at times may be considerable during humid periods in otherwise generally arid surroundings. This model is the one mostly applied to the evaporites formed in the Mediterranean during the Messinian (Hsü et al. 1978).

The third type, *deep basins under thick water cover* (C), necessitates a particularly intense rate of evaporation which would allow the supersaturation of the water column, crystallization of the salts, and their settling to the bottom. These deposits are usually rather regular, having formed in deep

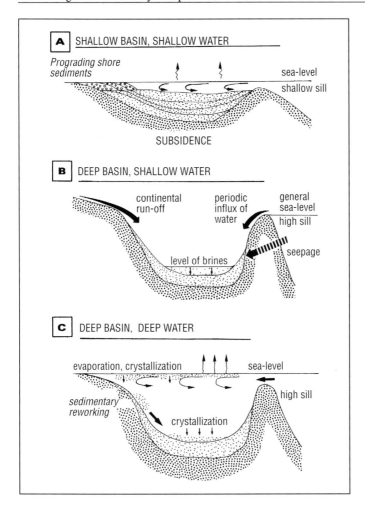

Fig. 1.25. Types of evaporitic basins (after Kendall, in Walker 1984)

water, and individual evaporite beds or laminae may frequently be followed over great distances due to the absence of wave action. They may be interrupted at times by resedimentation phenomena resulting from density and gravity currents issuing from the basin edge. The minerals assume shapes controlled by precipitation either at the water/air interface (fine-grained accumulations of halite cubic crystals) or at the water/sediment interface such as upward-growing halite chevrons, or large gypsum crystals with impurities and showing signs of dissolution. The majority of the large fossil evaporitic basins appears to belong to this type.

The evaporitic sequences are frequently characterized by the presence of *multiple sedimentary cycles* on the m-scale which indicate the alternation of arid and humid periods. Thus the upper Messinian of the Mediterranean is made up of alternations of gypsum, marls, and sands resulting from the succession of periods of chemical and detrital sedimentation. Laminations of 1–10-mm thickness are a common feature of these deposits. They are made up of alternations of dolomite/anhydrite with shale or shale/dolomite/anhydrite with halite, attributable to seasonal climatic fluctuations, and of alternations of halite with sylvite/carnallite resulting from periodic interruptions of the otherwise intense evaporation. The various structures are frequently deformed or destroyed by mechanical flow effects (*halokinesis*), and by processes of *recrystallization* and *dissolution* during the diagenetic evolution of the rather unstable evaporite minerals.

2 Properties of Sedimentary Particles

2.1 Grain Size

2.1.1 Scales

Sedimentary particles range from the fine dust transported by high-altitude winds to gigantic erratic blocks moved by glaciers. Due to their highly diverse shapes, during granulometric investigations the particles are conventionally considered as spheres of equivalent volume. This naturally will lead only to approximate results: it is especially the case for bioclastic sands with frequently curved particles which pass diagonally through square-mesh screens, and for the clay mineral platelets, the size of which is studied by sedimentation in reference to spherical grains (cf. Rivière 1977). The subdivision of sedimentary particles in three large categories, sand (2–0.063 mm), silt (0.063–0.004 mm), and clay (below 0.004 mm), leads to a general classification of sediments (Fig. 2.1). The term clay is used here without its mineralogical connotation. In

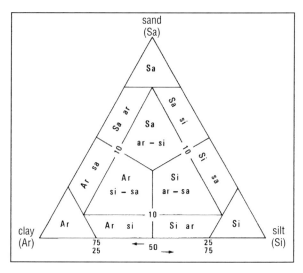

Fig. 2.1. Granulometric nomenclature of medium- to fine-grained sediments (*Si ar* argillaceous silt; *Ar si-sa* silty-sandy clay)

Table 2.1. Grain-size scales and classification.

US standard		Terminology			Afnor norm		Method of determination
mm	Ø				α	mm	
256	−8		Boulders		−24	250	Direct measurement
					−23	200	
			Cobbles		−22	160	
128	−7				−21	125	
					−20	100	
					−19	80	
64	−6				−18	63	
			Gravel		−17	50	
					−16	40	
32	−5				−15	31.5	
					−14	25	
		Rudites	Pebbles		−13	20	
16	−4				−12	16	
					−11	12.5	
					−10	10	
8	−3				−9	8	
					−8	6.3	
					−7	5	
4	−2				−6	4	
	{etc}		Granules		−5	3.15	
					−4	2.5	
2	−1				−3	2	
{1.68}	{−0.75}				−2	1.6	
{1.41}	{−0.5}			Very coarse	−1	1.25	
{1.19}	{−0.25}						
0	0				0	1	
{0.84}	{0.25}				1	0.8	
{0.71}	{0.5}			Coarse	2	0.63	
{0.59}	{0.75}						
0.5	1	Arenites	Sand	Medium	3	0.5	Sieving
					4	0.4	
	{etc}				5	0.315	
0.25	2				6	0.25	
				Fine	7	0.2	
					8	0.16	
0.125	3				9	0.125	
				Very fine	10	0.1	
					11	0.08	
0.0625	4				12	0.063	
				Coarse	13	0.050	
					14	0.040	
0.0312	5				15	0.031	
				Medium	16	0.025	
					17	0.020	
0.0156	6		Silt		18	0.016	
				Fine	19	0.0125	
					20	0.0100	
0.0078	7				21	0.008	
		Pelites, Mud		Very fine	22	0.0063	
					23	0.005	
0.0039	8				24	0.004	
			Clay		25	0.0031	Sedimentation
					26	0.0025	
0.0020	9				27	0.002	
					28	0.0016	
					29	0.00125	
0.00098	10				30	0.001	
					31	0.0008	
					32	0.00063	
0.00049	11				33	0.00050	

the majority of cases, the size scales are subdivided into *granulometric classes* on a geometric scale. The one most widely used is that of Udden-Wentworth (US Standard) in which the upper limit of each principal class is double the limit of the preceding class, the basic diameter corresponding to 1 mm (Table 2.1). This type of progression results in a logarithmic graphic presentation. In day-to-day use the latter is often replaced by an arithmetic system of better readability, in which each main class limit is given by a full digit: the progression of Udden-Wentworth thus corresponds to the Φ (phi)-scale of Krumbein in which $\Phi = \log 2$ mm and $\Phi_0 = 1$ mm.

The French granulometric system (AFNOR) corresponds to a geometric progression according to $\sqrt{10/10}$, with a logarithmic transformation through substitution by an arithmetic scale a ($a_0 = 1$ mm) as shown in Table 2.1.

2.1.2 Graphic Presentation of Grain Size Data

Grain size distributions can be illustrated in a number of ways (Fig. 2.2). The most simple one is the *histogram* or bar chart, in which the weight percentage of each granulometric class is expressed by a vertical bar. The discontinuous character of the histogram may be avoided by constructing the *frequency curve* which passes through the center of the bars and allows the definition of *modes* as well as of the bars themselves. The *cumulative frequency curve* with an arithmetic scale is widely used. This method of presentation is derived directly from grain size determination by sedimentation. It allows the measurement of the *median,* the size which divides the total sediment in two equal parts, as well as the graphic calculation of various distribution parameters:

Mean grain size:
This supplies a good impression of the spread of the granulometric curve of a sample:

$$\mu_z = \frac{\Phi_{16} + \Phi_{50} + \Phi_{84}}{3} \quad \text{(Folk 1966)}$$

(Φ_{16} = diameter of particles in Φ-units which corresponds to 16 weight – % of the sample)

Sorting:
This illustrates the size range of a sample compared to the mean size:

$$\Phi_i = \frac{\Phi_{84} + \Phi_{16}}{4} + \frac{\Phi_{95} + \Phi_5}{6,6}$$

0 – Φ – 0.35 very well sorted
0.35 – Φ – 0.50 well sorted
0.50 – Φ – 0.71 reasonably sorted
0.71 – Φ – 1.00 moderately sorted
1.00 – Φ – 2.00 poorly sorted
2.00 – Φ – 4.00 very poorly sorted

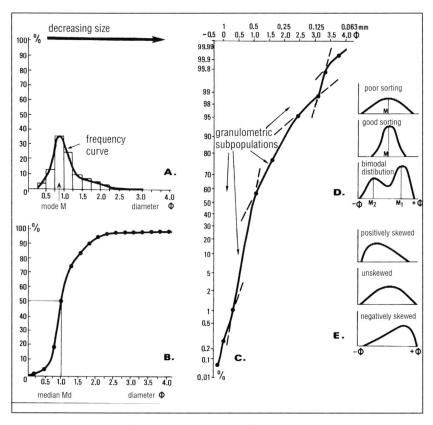

Fig. 2.2 A–E. Examples of graphic representations of grain-size parameters. **A** Histogram and frequency curve; **B** cumulative frequency curve on an arithmetic scale; **C** cumulative frequency curve on probability scale, showing different populations by Gaussian sections and allowing the distinction of various granulometric fractions; **D** sorting; **E** skewness

Skewness:
This illustrates the symmetry of a distribution, or the relation of the tails of the distribution to the mean. A preponderance of the fine fractions is indicated by positive values, that of the coarse fractions by negative values.

$$S_{Ki} = \frac{\Phi_{16} + \Phi_{84} - 2\Phi_{50}}{2(\Phi_{84} - \Phi_{16})} + \frac{\Phi_5 + \Phi_{95} - \Phi_{50}}{2(\Phi_{95} - \Phi_5)}$$

$+1.00 - S_{Ki} - +0.30$ strongly skewed to fine sizes
$+0.30 - S_{Ki} - +0.10$ skewed to fine sizes
$+0.10 - S_{Ki} - -0.10$ symmetrical, unskewed distribution
$-0.10 - S_{Ki} - -0.30$ skewed to coarse sizes
$-0.30 - S_{Ki} - -1.00$ strongly skewed to coarse sizes

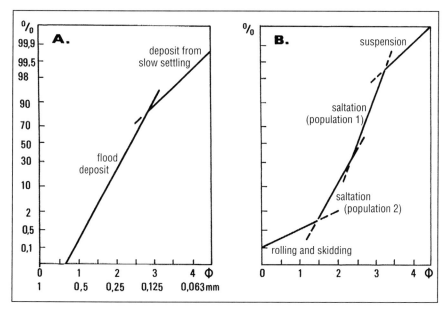

Fig. 2.3 A, B. Cumulative frequency curves on probability scale. **A** Fluvial sand; **B** marine beach sand (after Visher, in Friedman and Sanders 1978)

The construction of a double-logarithmic cumulative distribution curve in a probability plot permits the comparison of the distribution of a sample with a Gaussian curve, and thereby the identification of potentially present granulometric fractions introduced from different sources or by different depositional mechanisms, even in samples which in a semilogarithmic curve appear to be homogeneous, as shown in Fig. 2.2 (cf. Rivière 1977). That sediments deposited by rivers, or on beaches, may show different numbers and types of granulometric fractions, is the result of different mechanisms of transport and deposition in these two environments (Fig. 2.3).

2.1.3 Characterization of Sedimentary Environments

The comparison of granulometric curves and the parameters derived therefrom permits the characterization of ancient and recent environments of sedimentation, especially if the samples are collected from the same series of the same sedimentary level. As the distance from the source frequently is expressed by a decrease of the mean size, the mixture of autochthonous and allochthonous components will appear as a bimodal distribution (Fig. 2.2 D), whereas the passage from agitated to quiet zones is expressed by a modification of the cumulative curves and their respective parameters. In general terms, the *mean size* is the expression of the force of a current, water or wind,

52 Properties of Sedimentary Particles

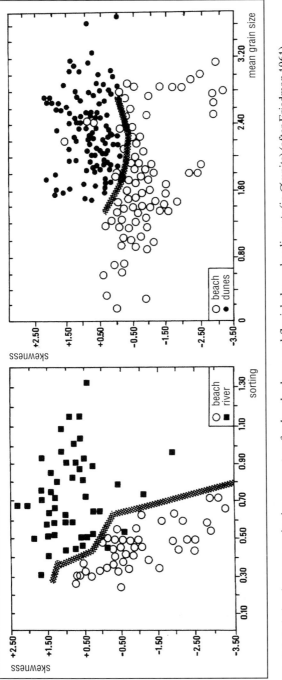

Fig. 2.4. Relation between grain-size parameters for beach, dune, and fluvial channel sediments (in ∅ units) (after Friedman 1961)

capable of bringing into motion a certain sediment. A sedimentary horizon in which the grain size decreases to the top, bears witness of a current slackening at a given point as time progresses (e.g., a diminishing turbidity current).

Sorting allows an appreciation of the size-grading processes active during transport and deposition. Thus, the winnowing of the fine particles on beaches will lead to well-sorted sands and gravels. Well-sorted aeolian deposits are depleted in the coarser fractions which are not brought into motion and remain behind as a lag deposit, as well as in the finer fractions which are carried further. In contrast to this, glacial deposits virtually always exhibit poor sorting of particles, resulting from the transport without separation, of all sizes broken from the rocky substrate.

Skewness is controlled more by the depositional processes than by the transport conditions. Thus certain fluvial deposits show a strong positive skewness, caused by the enrichment of a fine fraction settled out from a receding flood. The evaluation of the various granulometric factors facilitates the identification of ancient depositional environments (Fig. 2.4), when these factors are seen in their sedimentological context as widely as possible and when treated by appropriate statistical methods (Rivière 1977).

2.1.4 Influence of Transport Processes

The size reduction of particles during transport is the result of fragmentation and attrition under the influence of a number of poorly quantified processes. The impact of grains carried by the fluid plays an important role. This influence makes itself more felt during transport by air than by water, where it is counteracted by the effects of buoyancy and viscosity. Experience shows that over the same transport distance, the attrition of quartz in air is 100 to 1000 times more effective than in water. Quartz sands may travel over 1000 km in water without losing more than 0.1% by weight. The effects of attrition are considerably reduced for small particles and virtually disappear for sizes of less than 0.05 mm transported by wind or water (in Leeder 1982). This explains the weak rounding encountered in certain fine-grained marine beach sands mainly exposed to wave action (e.g., St. Maries-de-la-Mer, Camargue/France). Clearly, care must be taken when using grain rounding as a parameter in paleogeographic reconstructions.

2.2 Shape, Surface, and Orientation of Grains

2.2.1 Shape

One way of approaching an evaluation of the shape of sedimentary particles lies in their comparison with geometric standards: cylinders or rods

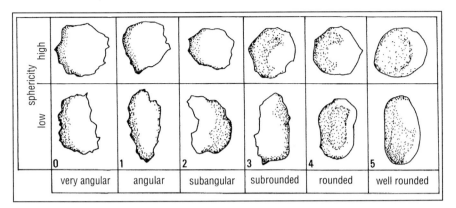

Fig. 2.5. Visual determination of sphericity and roundness (after Pettijohn et al., in Tucker 1982)

(length ≠ width = thickness), elongated plates with angular or rounded borders (l ≠ w ≠ t), square or disc-shaped plates (l = w ≠ t), cubes or spheres (l = w = t). The *sphericity* of a grain, estimated from the length-width ratio, denotes the degree at which a particular body approaches that of a sphere. *Roundness* may be estimated visually as a function of the sphericity (Fig. 2.5), or calculated from the ratio of the smallest diameter inscribed on the periphery of the grain to its length. Numerous morphometric applications have been developed, in particular by Cailleux and Tricart (1959). The final shape depends not only on the shape and hardness of the source material, but also on the mode and duration of transport. In fluvial systems, quartz pebbles which are harder than those of the frequently used limestone, but less hard and breakable than those of chert or flint, allow a good assessment of the mechanical shaping processes during long-distance transport.

2.2.2 Surface Morphology

This has been the subject of direct observation for a long time already: pebbles striated and indented by glacial transport, frosted or surface-sculptured by the bombardment from wind-transported particles, etc. Studies under the optical microscope (*morphoscopy*), one of the classical methods, permit the recognition of various processes such as on grains resulting from physical weathering, from glacial or fluvial transport, lustrous grains shaped by long transport in water (marine beaches and beaches of large lakes, lower reaches of large rivers), or rounded grains frosted by transport in air. Grains may also be subjected to several processes either directly or through reworking (Cailleux and Tricart 1959). Recently, observations under the transmission electron microscope (*exoscopy*) have been introduced (cf. Le Ribault 1977). Mostly applied to *quartz,* these methods help in the characterization of the sedimentary environment, the mode of transport, and the relative energy level through the identification of chemical and physical phenomena which affected the grain surfaces:

glaciers:	conchoidal fractures, no wear
continental soils:	frequently, formation of a thin coating of amorphous silica
rivers:	slight wear through attrition, dissolution of the siliceous coating acquired in the soils
beach, intertidal zone:	on emersion, deposition of silica on the surface depressions of grain through evaporation; on immersion, partial dissolution and mechanical attrition of the surface layers;
shallow marine bottoms:	superficial solution due to undersaturation of sea-water in silica, polishing through attrition, V-shaped impact marks;
dunes:	numerous impact marks of grains in non-viscous environment, resulting in a polished appearance and rounded shape.

The application of these techniques permits interesting conclusions, especially in the reconstruction of continental and coastal environments, on the basis of cycles of reworking of quartz grains and on the relative importance of mechanical and chemical processes. This requires a careful statistical approach due to the tendency of quartz to be subjected to sedimentary reworking and to the possibility of diagenetic modifications such as dissolution or recrystallization.

2.2.3 Spatial Arrangements

This is the result of the mode of accumulation (Fig. 2.6), often reflecting the depositional conditions and controlling certain physical properties of the resulting rocks. An *isotropic fabric,* consisting of grains arranged without preferred orientation, frequently results from very rapid mass deposition (turbidity currents, grain flows). Transport may also take place very slowly, within a denser mass, such as in mud flows or basal moraines. *Anisotropic fabrics* result from unidirectional currents: glacial bottom gravels oriented parallel to the direction of transport and inclined at 20°–40° in the upflow direction; fluviatile pebbles and sand rolled on the bottom in parallel to the flood current or moved by saltation parallel to the flow during normal discharge and plunging at 15°–30° in the upstream direction; littoral pebbles imbricated with shallow upstream plunge of about 15°, thereby offering minimum resistance (Fig. 3.27A).

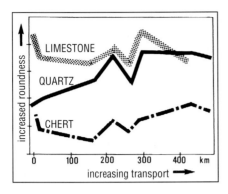

Fig. 2.6. Evolution of roundness for pebbles of different materials in the lower section of Colorado River, Texas (after Sneed and Folk, in Friedman and Sanders 1978)

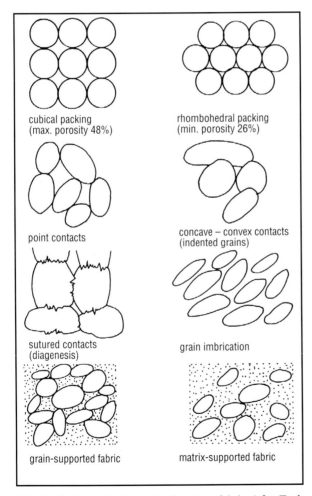

Fig. 2.7. Grain orientation and sedimentary fabrics (after Tucker 1982)

Porosity, which represents the open spaces in a given sedimentary volume, always depends on the type of packing. There are all intermediate stages between a cubical packing of high porosity and a more rhombohedral one, in which the open spaces are only small (Fig. 2.7). Shelly sands consisting of heterogeneous particles tend to possess particularly high porosities. Rapidly sedimented clays frequently have porosities in excess of 90% because of the flocculated nature of the clay flakes. Compaction, migration of pore water, or vibration strongly reduce the porosity of clays. Slow deposition such as settling of independent particles, or extended hydrodynamic influences such as experienced by shallow-marine sands due to the action of waves and tides, favor the establishment of an optimum grain arrangement and thus of reduced porosities. Diagenesis generally also leads to lower porosities through compression, dissolution, and recrystallization. *Permeability*, the ability of fluids (water, hydrocarbons) to migrate within a rock, frequently is independent of the porosity. Whereas the voids between the individual clay minerals, although numerous, are too small to allow a notable permeability in clayey rocks, the pores and joints in sandy rocks and certain limestones result in the good reservoir characteristics of such rocks.

2.3 Classification of Sediments

Knowledge of the various characteristics of a sediment permits the assessment of its state of evolution, its maturity, and also a better understanding of its history. An immature sediment generally has a wide granulometric range, abundant matrix, and will exhibit poor sorting, angular grains as well as high porosity and permeability. Such a sediment is the result of rough hydrodynamic actions, slow or weak, as encountered in certain fluvial or glacial environments and during marine resedimentation. The opposite will be the case for a mature sediment, which may be used as evidence for active and prolonged hydrodynamic processes in water or air, such as in littoral or desert dunes, beaches, and other shallow-marine exposed environments.

The detailed description of a sediment will also allow its precise classification in combination with data on the grain size (Fig. 2.1), as well as on the overall nature of the various constituents (e.g., limestone vs clay) and the individual composition (e.g., main components). In literature, various classifications of a descriptive, genetic, or general nature have been proposed (e.g., Vatan 1967; Friedman and Sanders 1978). There is no simple classification which will take into account all the main characteristics of a sediment. An exemplary method has been developed by Dean et al. (1985) for open-marine sediments collected during the Deep Sea Drilling Project. It leads to a somewhat strained nomenclature which, however, is concise, ordered, of reasonable precision, rapidly established, and easily understood (Fig. 2.8). Being entirely descriptive, it eliminates all genetic connotations such as detrital ooze or pelagic clay. It is based on the three main compounds of marine

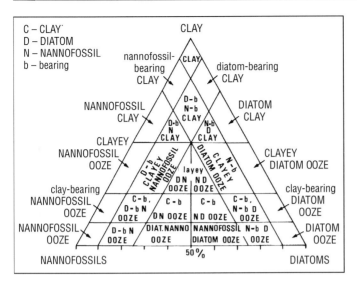

Fig. 2.8. Classification of oceanic sediments according to main constituents, i.e., clay, diatoms, and nannofossils (after Dean et al. 1985). For explanation, see text

sediments, namely non-biogenic, biocalcareous, and biosiliceous matter, using certain quantity thresholds as estimated from studies of polished or thin sections. There are four basic steps in the application of this classification:

1. The designation of the first order covers the sedimentary constituent present in a quantity of 50% or more. In the case of non-biogenic material, it reflects the particle size (clay, silt, or sand). In the case of biogenic material, it is replaced by the term "ooze", modified by the nature of the organisms: nannofossil ooze, diatom ooze.

2. The second order relates to compounds present in quantities between 25%–50% which modify the first-order terms:
non-biogenic: clayey, silty, sandy;
biocalcareous: nannofossil, foraminifer, or calcareous;
biosiliceous: diatom, radiolarian, or siliceous.

3. The third-oder concerns compounds present in quantities between 10%–25%, and further modifies the first-order terms with the aid of the suffix "-bearing": clay-bearing, diatom-bearing.

4. Constituents present in quantities below 10% are not expressed in the final term which contains, read from left or right in increasing order of abundance, the various compounds: e.g., foraminifer-bearing, clayey nannofossil ooze.

A few examples with data obtained from microscopic estimations illustrate the method:

60% nannofossils, 30% foraminifera, 5% radiolarians, 5% clay: foraminifer nannofossil ooze:

Table 2.2. Induration stages of sedimentary deposits (after Dean et al. 1985)

Main component	Soft	Friable or fissile	Hard
Bio-calcareous	Calcareous ooze	Chalk	Limestone
Bio-siliceous	Siliceous ooze	Diatomite Radiolarite Lussatite	Porcellanite Radiolarite Chert
Non-biogenic	Clay Silt Sand	Clay-shale Silt-shale Friable sand	Claystone Siltstone Sandstone

40% nannofossils, 35% foraminifera, 15% radiolarians, 10% clay: clay- and radiolarian-bearing foraminifer nannofossil ooze;
35% diatoms, 35% clay, 30% radiolarians: radiolarian clayey diatom ooze;
50% sand, 30% silt, 10% foraminifera, 5% radiolarians, 5% clay: foraminifer-bearing silty sand.

The finding of the exact term is carried out in three main steps: estimation of proportions of biogenic and non-biogenic constituents, proportion of compounds within each of the two groups and, finally, the ranging of the various constituents according to the thresholds (10%, 25%, 50%). The resulting nomenclature permits a great flexibility in application. It may be adapted to sediments consisting of rare, but locally abundant materials (volcanic glass, metal oxides, zeolites), or to sediments with poorly characterized components (undifferentiated carbonates = calcareous; recrystallized silica = siliceous). Furthermore, it provides for an indication of the degree of consolidation of the respective sediments by introducing similar rules for their various main constituents (Table 2.2). Such a classification will not exclude widely used classic terms such as marl, mudstone, etc., but may be applied wherever precision is required for a better description and understanding of a sedimentary sequence.

3 Deposition of Sediments

3.1 Principal Transport Mechanisms

3.1.1 Sedimentary Particles

Movement of the carrying fluids water and air and the solid particles suspended in them, is effectively controlled by density, viscosity, type of discharge, its strength and velocity, as well as by size and shape of the particles and their surface properties, all factors which may be readily quantified (in Gibbs 1974; Leeder 1982; Allen 1984). The *velocity of the current* and the *size of the particles* in it constitute the fundamental parameters controlling the start of movement and relative transport through erosional and sedimenta-

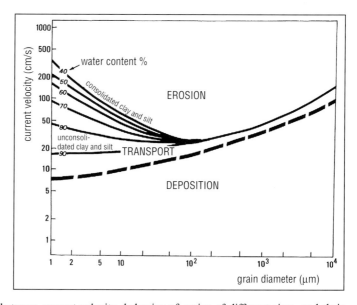

Fig. 3.1. Relation between current velocity, behavior of grains of different sizes, and their degree of consolidation (diagram of Hjulström, modified by Postma and Gardner, in Blatt et al. 1980)

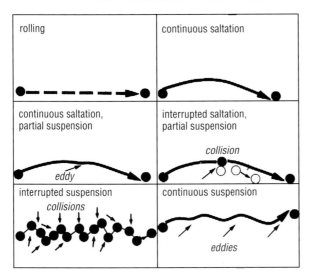

Fig. 3.2. Principal modes of transport of individual particles (after Leeder 1979, in Leeder 1982). The position of the bottom is indicated by the *horizontal lines*

tional phenomena (Fig. 3.1). The remarkable resistance of clayey and silty rocks against erosion, especially when compared to sandy rocks, is the result of their electrolytic properties, occurring as flocculates and aggregates, and increases with the degree of cohesion. When a current is strong enough to erode a fine-grained deposit, clay will rather be detached in the form of aggregates than as isolated flakes.

Once the threshold of mobility is reached, the particles may be subjected to three continuous or interrupted modes of transportation (Fig. 3.2). *Rolling* implies the continuous contact of the moving particles with the floor and at times includes sliding especially of more platy particles. During *saltation* the grains abruptly leave the bottom generally at large angles above 45°, and thereafter fall back slowly in more or less continuous fashion. Grains in *suspension* are completely surrounded by water, frequently close to the bottom, and move usually in long irregular trajectories. To keep in suspension the grains which are denser than water, upward movement is required. This results from turbulences in the carrying fluid, which need to be stronger in the case of larger and less buoyant particles. The poor carrying capacity of air for transport by saltation or in suspension, resulting from its very low densities and viscosities, is compensated by the frequently very high velocities, and by the amplitude of upward movements resulting from impacts. Differences in the behavior of the particles, moving either close to the bottom or farther away from it, necessitate a distinction between two particular types of loads, viz. bedload or traction load, and suspension load (in Leeder 1982).

Close to the bottom, currents frequently result in renewed uptake of sediment, thereby increasing the turbidity of the fluid. This is the case with mud

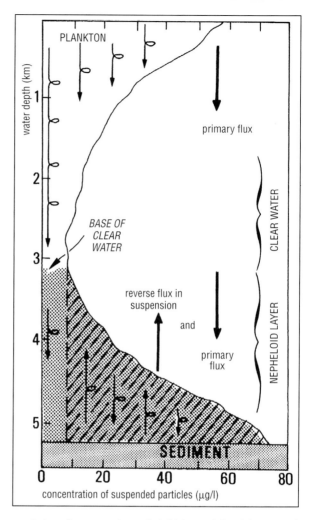

Fig. 3.3. Formation and characteristics of an oceanic nepheloid layer (after Biscaye and Eitreim 1977)

"plugs" in estuaries marked by partial discharge during low tide. It also occurs in the oceanic environment in the *nepheloid bottom layers* where, under the influence of lateral or longitudinal currents, previously deposited sedimentary matter may be again taken up. Particle concentrations close to the bottom may be comparable to those in surface waters resulting from intense planktonic productivity. Between these two layers of "charged" water there is a clear zone which due to dissolution and oxidation contains little matter coming from the surface zone, and is depleted in particles derived from the bottom (Fig. 3.3). The rate of flux of the particles derived from the overlying water and from the sedimentary bottom permits a quantification of the residence time of the particles in a nepheloid layer. In the Atlantic the times are rather short, in the order of one week within 15 m from the bottom, and months in the first 100 m of the bottom zone.

3.1.2 Flows of Sedimentary Particles

The massive transport of particles may be viewed from various aspects such as slope and available energy, and especially the type of relation between the grains, their matrix, and the carrying liquid. The remobilization of a fine-grained, poorly consolidated deposit, may take place without much dislocation of particles by slumping, which consists of a rotational *sliding* within a pasty mass. The sediment is deformed in the greater part by folding or synsedimentary faulting (cf. Sect. 3.4.2); however, it will lose its cohesion only locally.

Under the influence of gravity, true flow takes place by four main processes with various intermediate stages. *Grain flows* are dense masses in granular flow in which the particles collide with, and dislocate each other (Fig. 3.4). Water is either absent, as in sandy avalanchites, or not strong enough to reduce the forces of friction. Movement may only be initiated when the slopes are larger than the natural angle of stability in subaerial or subaqueous environments. During the movement, inverse-graded bedding can develop through differential friction close to the bottom, or through downward migration of the finer particles between the coarser ones. *Debris flows* consist of particles of a wide size range, from silt to cobbles, carried by a silty-clayey aqueous matrix often of large proportion. Such masses, behaving like a paste with lumps, can originate subaerially as well as subaqueously on rather gentle slopes, and flow over large distances without becoming appreciably sorted. In *liquified flows* the grains are displaced and dispersed as dense suspended matter in water, and result from dissociation of aggregates through impacts. The grains are maintained in suspension by the moving water.

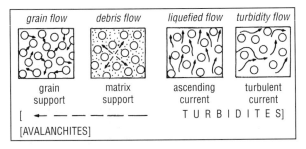

Fig. 3.4. Principal types of gravity flows of sedimentary particles with examples of resulting deposits

Table 3.1 Main types of mechanisms of transportation, and possible connections

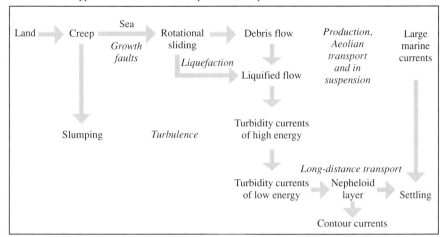

In motion, the mass has the consistency of a homogeneous liquid paste suffering only weak friction. As long as the carrying liquid cannot escape to the top, such flows are able to move on very low-angle slopes. *Turbidity flows* are characterized by the suspension of particles in a turbulently flowing liquid. They may move over exceedingly long distances on surfaces with extremely low or virtually no inclination. Such a flow represents a dense liquid mass within which horizontal and vertical sorting is taking place as a function of the velocity of the current and the mass itself, as well as of the grain-size distribution of the particles concerned (cf. Sect. 3.3).

In general, the mechanisms described above are encountered especially in the submarine environment off exposed zones submitted to other gravity movements such as block falls, and slow ground movements like solifluxion or soil creep. Slumping mainly characterizes the proximal subaqueous zones below points of sediment input such as deltaic lobes, detrital fans during progradation, etc. grading in the downstream direction into debris flows, liquified flows, or turbidity flows of various energy levels (Table 3.1).

The mutual continuity of these gravity phenomena tends to be replaced in the direction of the open sea by slow movements conducive to sedimentation by settling, as well as by basic transport mechanisms leading to contour currents deeper down (cf. Sect. 3.4.4). Gravitational mass flows are always prone to occur suddenly in areas of notable submarine relief and in thick sedimentary accumulations.

3.2 Formation of Basic Sedimentary Structures

3.2.1 Introduction

During deposition, the sedimentary particles arrange themselves under the influence of the same hydrodynamic and aerodynamic forces that are responsible for transport to the place of deposition. This arrangement results mostly in a large variety of structures controlled by the variability of the respective forces. This explains why the diversity of sedimentary structures is most pronounced in coastal environments and along the continental shelf. Due to their moderate grain size and weak cohesion, as well as the great durability of the individual grains, quartz sands are particularly suitable for the formation and conservation of sedimentary structures (e.g., Scholle and Spearing 1982). However, sedimentary structures are also common in bioclastic carbonate series (e.g., Scholle et al. 1983), in volcaniclastic rocks, and in resedimented evaporites. They are sometimes also formed in fluid plutonic magmas during crystallization. Although clay and gravel deposits are less favorable for the formation of sedimentary structures, they also show different types of structures in many cases characteristic of the mode of deposition (in Collinson and Thompson 1982). Fine-grained deposits may reveal their sedimentary structures only through microscopic investigation (cf. Stow and Piper 1984).

A lack of sedimentary structures is characteristic of particulate deposits resulting from continuous settling or from gravitational mass flows deposited under violent flow conditions. It may also result from obliteration of earlier structures by bioturbation or diagenetic processes.

A huge amount of published experimental data and results of detailed studies on sedimentary structures appeared in the last decades, starting with the synthetic publication of Pettijohn and Potter (1964). More recent references comprise detailed photographic treatises, as well as manuals on practical application (e.g., Reineck and Singh 1980; Collinson and Thompson 1982). Other works cover theoretical, experimental, and geological aspects of the subject (Allen 1984). We shall limit ourselves in the following mainly to practical considerations of the identification of sedimentary structures, their succession, and their hydrodynamic and geological significance. Many sedimentary structures are developed at the top or the base of a bed, either as direct imprints (marks) or as casts originally filling marks. Repetitive sedimentary structures, either complex in nature or organized in sequences, can be observed in sections of the respective beds.

3.2.2 Formation of Ripples, Current Structures

3.2.2.1 Flow Regimes

Formation, size, and shape of current marks in the aqueous environment depend mainly on the flow regime of the fluid, and on the size of the sand

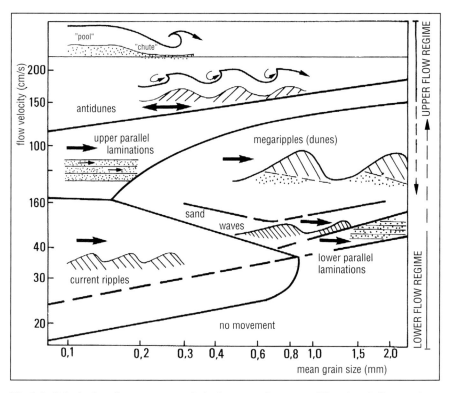

Fig. 3.5. Principal sedimentary morphologies according to unidirectional flow regimes (modified after Blatt et al. 1980; Reineck and Singh 1980); *thin arrows* direction of fluid flow; *thick arrows* direction of particle flow. The size of the sedimentary structures is not to scale

particles. Two different regimes of flow are distinguished (Fig. 3.5). In the *lower flow regime* the resistance of the particles against movement is high and their rate of displacement only moderate. The undulations of the water surface are not in phase with those of the sediment surface. Strong size sorting may suddenly occur and the finest particles are preferentially washed away. Under very low current velocities, the sediment surface is flat and undeformed, and no transport of sedimentary particles occurs.

With increasing velocity and turbulence, small asymmetric ripples start to form which may grow in size to form megaripples, dunes, or sand waves. In the *upper flow regime* the particles are intensely displaced in a continuous layer over the sedimentary bottom, virtually independent of their size. Undulations of water surface and sedimentary bottom are in phase with each other. With further increases of current velocity, planar bedding starts to form, followed by more or less symmetrical antidunes, and eventually by chutes and pools, surface marks which are alternatively highly deformed or subhorizontal.

3.2.2.2 Formation of Asymmetric Ripples

Ripples are the fundamental expression of the movement of fluids over a sedimentary surface. They result from the arrangement of the sedimentary particles under the influence of water or wind, and are oriented perpendicularly to the current responsible for their formation. The most common form, asymmetric ripples, results from a *unidirectional current* (river, tide, density current, wind, etc.). They are characterized by a rather low upstream slope and a more inclined downstream slope. Sedimentation of the moving sand takes place in the form of successive micro-avalanches on the downstream side. This results in small repetitive oblique laminations, and an advance of the ripple in the direction of the current (Fig. 3.6). When the depth of the overlying water is small, tending to emersion, the ripples become irregular, start to bifurcate, and to develop flattened crests. Asymmetric wind ripples usually exhibit sharp crests, elevated ripple indices (ratio of height to length), and numerous bifurcations of their crests.

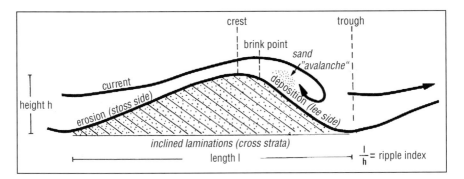

Fig. 3.6. Formation and characteristics of an asymmetrical ripple

3.2.2.3 Succession of Current Structures Under Increasing Energy

In Fig. 3.5 the succession of the current structures resulting from an increase in energy level is shown schematically. The *lower plane beds,* without movement of particles, are characteristic of sedimentary bottoms over which the currents are only weak, as in environments of slow settling at any depth and in cohesive argillaceous deposits. The initial movement of particles leads to small accumulations of sand behind larger grains. Ripples start to appear at velocities of about 14–25 cm/s, with sand grains in the size range 0.1–0.6 mm (in Blatt et al. 1980).

The typical *small ripples* exhibit 10–30 cm mean wavelength (maximum 60 cm) and 0.3–6 cm height, characteristic of sandy-silty sediments with an upper grain size of 0.6 mm (Reineck and Singh 1980). They are developed in most continental and marine environments, irrespective of the water depth. In coastal environments their appearance is controlled directly by the current

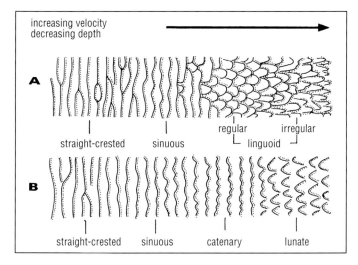

Fig. 3.7 A, B. Surface expression of crests of small (*A*) and large (*B*) sedimentary ripples (several cm and several dm resp.) as a function of current velocity and depth of water cover (after Allen in Fairbridge and Bourgeois 1978). *Dots* points of deposition in downstream direction

velocity and the depth of the water cover (Fig. 3.7). The crests of the ripples are arranged in a rectilinear pattern under slow currents, becoming gradually more sinuous and even discontinuous and linguoid (convex side pointing in the downstream direction) as the energy level increases (Fig. 3.7). Due to their asymmetry, current ripples are excellent indicators of polarity of movement. They may also be found in interference patterns of two different sets, superimposed over each other at various angles, resulting from two successive, differently oriented current patterns. Such cross-ripples, formed by currents of comparable or different strength, are generally characteristic of very shallow environments where they result from the refraction of different types of waves, or the succession of one wind direction by another one. When the water cover becomes extremely thin (1–2 cm) and the energy level remains high, the ripples take a rhombohedral shape, typical of the exposed shore environments in tidal seas such as beaches battered by waves and the inner side of slightly emerged sand bars covered by large waves during washover. As time progresses, ripple fields become superimposed onto each other through lateral downstream migration and are called climbing ripples.

Sand waves start to develop with growing current force from plane beds or small ripples of medium-grained particles (Fig. 3.5). The resulting large sedimentary structures or bed forms are generally at right angles to the current and are easily recognized in bathymetric profiles or side-scan sonar surveys (e.g., Berné et al. 1986). Their mostly rectilinear crests are several meters to several hundreds of meters long, at heights of up to several meters. The larger sand waves are also known as giant ripples. They sometimes occur in groups, forming large sand banks elongated in the direction of the current on continental shelves subjected to tidal currents such as the English Channel and the southern part of the North Sea. Deep-sea sedimentary drifts made up of finer-grained sediments proceed under similar hydrodynamic processes. Consisting of material larger than 0.25 mm, and built up by currents of 30–80 cm/s velocity, sand waves harbor smaller ripples on their upstream side which tend to advance in the current direction through successive avalanches.

Megaripples or dunes form subaqueously at mean velocities above those leading to sand waves, i.e., above about 60 cm/s. Their mean crest length is 0.5–10 m at heights of 0.06–1.5 m, and they consist of grains coarser than 0.1 mm. Their mode of formation and their internal structures are similar to those of normal ripples. In individually developed normal ripples, and megaripples, the inclined laminae deposited by avalanches in the downstream direction (foreset laminae) advance over a basal layer (bottomset laminae) resulting from settling on the downstream side of the crest. They are covered by the stoss-side laminae, a layer of fine-grained particles moved over the upstream side (Fig. 3.8). The shape of the megaripples varies with the speed of the respective current, but in a manner different from that of the smaller ripples (Fig. 3.7): the initially rectilinear crests become progressively more sinuous, then catenary, and eventually interrupted, presenting themselves as a sequence of lunate bodies with the convex side pointing in the upstream direction. Small ripples may develop on the megaripples, but tend to disappear as the energy level increases. The megaripples are frequently developed in high-energy, shallow-water environments such as rivers, estuaries, and shores exposed to waves, but are also found at greater depths, provided the currents are rather violent as in the English Channel. However, the nomenclature of megaripples, sand waves, and submarine dunes is still the subject of discussions as the mechanisms responsible for the formation of these structures are superimposed onto each other in a complex manner and at the same time are only imperfectly understood (cf. in Blatt et al. 1980; Reineck and Singh 1980). Whereas the recognition of these structures in recent environments is fairly well established, their identification and distinction in ancient sedimentary series is rendered difficult due to their large size in relation to the size of the

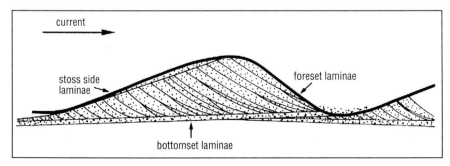

Fig. 3.8. Detailed structure of an asymmetric ripple or megaripple (after Reineck and Singh 1980). For details see text

available outcrops, and also due to the difficulty of conservation because of the erosion of the crests.

The *upper plane beds* are characteristic of the change-over from the lower to the upper flow regime. Where the surface is little deformed, the current is too rapid to permit the formation of megaripples, but not yet rapid enough to lead to the formation of antidunes. Particles are moved by traction, small jumps, and sliding over the bottom, thereby leading to longitudinal striations and various erosion or impact marks (cf. Sect. 3.3). The gradation of megaripples into the upper plane beds takes place more easily in the case of fine than of coarser sands.

Antidunes are ripple-like structures of usually symmetrical shape, the wavelength of which is in phase with that of the violent undulating current forming this bed form. Their length varies between a few centimeters to several meters, whereas their height is usually less than 50 cm. They are made up of particles of all sizes and forms in shallow water subjected to violent currents as in tidal channels. They represent rather unstable structures which may grow in height until breaching the water level in the form of breaking antidunes. They can also gradually disappear in standing waves. They can remain immobile or migrate in the downstream or upstream direction. In the latter case, which is the distinguishing factor, they collapse because of intense turbulent movement. Due to the interaction of erosional and depositional processes, they show little internal structure. Deposition takes place mainly in the form of thin regular laminae over the whole surface of the antidune (Fig. 3.9 A) or in layers on its upstream side in case of upstream migration (Fig. 3.9 B). Deposition on the downstream side is less frequent (Fig. 3.9 C).

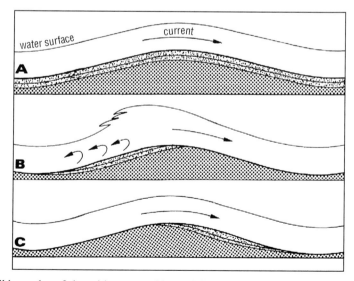

Fig. 3.9 A–C. Possible modes of deposition on antidunes (after Kennedy, in Reineck and Singh 1980). In **B** the sense of sediment migration is opposed to the current direction

Chute and pool structures are the result of extreme energy levels which lead to an irregular and discontinuous interaction of erosion, transport, and deposition. Zones of intense and rapid runoff, with slow but massive displacement of material and upstream movement, alternate with quiet areas exposed to lower flow regimes. Such structures are observed at present in certain sloped channels and water courses with elevated rates of sediment transport, but are only rarely encountered in ancient sediments.

3.2.2.4 Wave Action

Symmetrical ripples, or *wave ripples,* are formed under water through oscillating wave movements. They are usually observed on beaches and under shallow water cover. Their crests are mostly rather sharp (wavelength 1–200 cm, height 0.3–25 cm) and the troughs rounded. Transient micro-avalanches of sedimentary material occur on either side of the ripples as the waves advance or retreat. This results in inclined symmetrical, or chevron-shaped, laminae which are rounded over the crest of the ripple and approximately parallel to the surface of the ripple field (Fig. 3.12). The rounded crests often form as a consequence of wind action after emersion. When the two currents are not of equal intensity, a certain asymmetry may be observed in wave ripples. There is a distinction between current ripples exhibiting very little asymmetry, those with mostly rectilinear but frequently bifurcating crests (Fig. 3.10), and those without inclined microstratification. This distinction is of importance for ancient formations, as the current ripples will indicate the direction of the sedimentary transport, whereas the wave ripples only show its general orientation. As it is the case for current ripples, the succession of wave ripples exhibits vertical stacking in the form of climbing ripples (cf. Sect. 3.2.2.3).

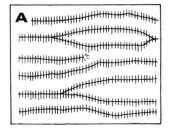 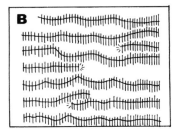

Fig. 3.10 A, B. Distinction between asymmetric wave ripples (**A**) and current ripples (**B**) on the surface of sediment bars (after Reineck and Singh 1980)

3.2.2.5 Distribution of Ripples in Different Depositional Environments

The formation of ripples is mainly controlled by the hydraulic conditions and the grain size of the sediment. The identification of the different types of ripples and the estimation of their relative frequencies are of great value in the reconstruction of past depositional environments and their hydrodynamic flow conditions (Table 3.2).

Table 3.2. Nature and relative abundance of various types of sedimentary ripples in main depositional environments (after Reineck and Singh 1980)

	Current ripples	Rhomboedral ripples	Climbing ripples	Megaripples, sand waves	Antidunes	Wave ripples
Rivers	++	–	++	++	–	–
Lakes	–	– –	– –	– –	– –	+
Lacustrine beaches	+	–	–	–	–	++
Coastal lagoons	–	–	– –	– –	– –	+
Intertidal zone	++	–	–	+	–	++
Tidal channels	++	– –	–	++	+	– –
Upper infratidal zone (lower shoreface)	+	– –	– –	+	– –	+
Lower infratidal zone (lower shoreface)	–	– –	– –	– –	– –	+
Transition zone	–	– –	– –	– –	– –	+
Sandy shelf	+	– –	– –	+	– –	–
Silty-clayey shelf	– –	– –	– –	– –	– –	–
Shelf edge and slope (without current action)	–	– –	– –	– –	– –	– –
Zones of gravity and contour currents	+	– –	+	–	–	– –
Deep basins	+	– –	– –	–	– –	?
Seamounts	+	– –	– –	–	– –	–

++ : abundant + : common – : rare – – : absent

3.2.2.6 Vertical Succession

As mentioned above, sedimentary ripples occur as *beds* of variable thickness but usually more than 1 cm. *Laminae* consist of inclined or horizontal layers in the millimeter range (e.g., Figs. 3.5, 3.8). To become fossilized, the ripples have to be buried, usually through the action of a subsequent current (e.g., river flood, or successive tides). Before depositing its load on the bottom, a current will frequently erode previously deposited material. This leads to more or less truncated ripples which are made up of sandy beds with thin inclined laminations. The successive variations in velocity and direction of the currents, frequently result in oblique truncations of the underlying ripples, leading to the formation of *cross-bedding* in which the individual units re-inclined against each other (Fig. 3.11 A). As the top of a truncated layer is always older than the overlying bed, and as the cross-stratifications are inclined in the downstream direction of the flow, the succession of current ripples in a sedimentary section furnishes excellent indications of the younging direction of the beds and of the direction of the currents that deposited the rock. This has been well established in innumerable sedimentary formations of aeolian, fluvial, deltaic, and marine origin (in Scholle and Spearing 1982). Various types of cross-bedding are encountered, depending on the relative sense of inclination of the individual laminae and of the beds themselves: subparallel or

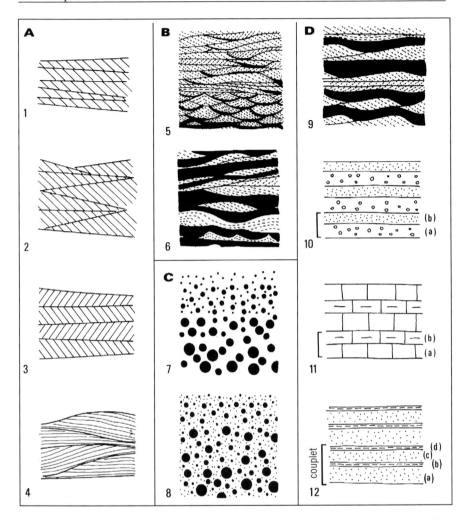

Fig. 3.11 A–D. Examples of vertical arrangement of basic sedimentary structures (partly after Reineck and Singh 1980). The scale varies according to energy, rate of supply, and grain size. **A** Inclined bedding: *1* subparallel (with cross-laminations); *2* inclined and crossed (crossbedding); *3* herring-bone cross-bedding; *4* hummocky cross-bedding. **B** bedding with episodic intercalations: *5* fine-grained material (*in black*) in coarser sediment (flaser bedding); *6* coarse-grained particles in finer sediment (lenticular bedding). **C** normal size-graded bedding: *7* in strict sense, without fine-grained admixtures at base; *8* in wider sense, with ubiquitous fine-grained matrix (N.B.: the same situation is developed in reverse gradedbedding). **D** alternations: *9* irregular alternation of undulating coarse and fine sediment (wavy bedding); *10* alternation of flood (*a*) or ebb (*b*) deposits; *11* alternation of limestone beds (*a*) and marls (*b*); *12* alternation in subtidal environments of deposits resulting from flood stage (*a*), high sea-level stage (*b*), ebb stage (*c*), and low sea-level stage (*d*) (double mud drape)

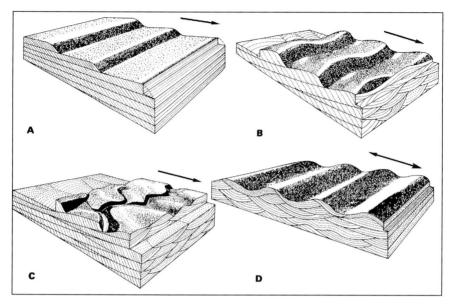

Fig. 3.12 A–D. Three-dimensional view of structures resulting from a sequence of current ripples with straight (**A**), sinuous (**B**), and linguoid crests (**C**), as well as from slightly asymmetric wave ripples (**D**) (after Reineck and Singh 1980). *Arrows* direction of current. The portions removed at the *right-hand edge* of the block diagrams permit a view of the horizontal section

oblique stratification, herringbone cross-bedding, hummocky cross-bedding, etc. (Fig. 3.11 A). The three-dimensional analysis of the structures developed in the various types of modern environments of current and wave ripples (Fig. 3.12), is of great help in the detailed reconstruction of the various structures and the hydrodynamic forces responsible for them in fossil environments.

The sedimentary beds deposited in succession may repeat themselves as long as the hydrodynamic conditions remain the same, thereby forming sets of beds. However, frequently the energy fluctuates with time. When the current responsible for the formation and migration is interrupted, like during high tide over estuarine sand banks, a thin, mm- to cm-thick layer of silty-clayey sediment can settle out to cover the top of the ripples, giving rise to *flaser-bedding*. Conversely, when an environment, in which usually fine-grained material is deposited, is invaded episodically by a sand-bearing current, this material can be laid down in the form of lenses or irregular stringers in the mud, resulting in *lenticular-bedding* (Fig. 3.11 B). Other variations of energy affect the deposits in a more regular manner, as during flood or low-water stages of rivers, or during the different stages of tides in the subtidal environment. This leads to an alternation of fine- and coarse-grained layers (Fig. 3.11 D), evidence of changes in the hydrodynamic regime and, in many cases, exact geological time markers (cf. Einsele and Seilacher 1982). Finally, the energy level within a bed may vary very gradually, leading to a progressive change in the degree of granulometric sorting, a phenomenon called *graded-bedding*. This is developed during the slackening of a turbidity flow, producing sorting of the particles in such a way that the coarser fractions are deposited in the lower part of the bed resulting in a fining-upward structure. When dealing with debris flows or certain volcanic products forming dense flows (ignimbrites), size grading is only imperfectly developed (Fig. 3.11 C). During the vertical growth of sand bars or ripples approaching the water surface, and also in active channels migrating laterally in a zone of slow settling sedimentation, an increase in the energy level can lead to inverted size grading or a coarsening-upward structure. The various changes of the energy level of an environment may be reconstructed from the study of the resulting sedimentary structures.

3.3 Formation of Post-Depositional Sedimentary Structures

3.3.1 Erosion Structures (Fig. 3.13)

3.3.1.1 Scour Marks

These are structures eroded into a finer-grained substrate by a turbulent current loaded with suspended particles. The fossilization of erosion marks is the result of the subsequent deposition of sand which after lithification is preserved as a cast in a mould. This explains why scour marks are observed as sole casts mostly in alternations of sandy beds with siltstones or mudstones.

Flute casts, sequences of aligned half-cones with the round thicker end pointing in the up-current direction, result from the erosion of sediment through cavitation. Although varying in length from 5–50 cm, flute casts on individual bedding planes are usually fairly uniform in length. They are found in deposits of rather contrasting depths, and are sometimes cut by transverse current structure, the so-called transverse scour casts indicative of shearing within a fluid (in Reineck and Singh 1980). Longitudinal furrows, ribbons, and ridges developing parallel to the current direction, result from more defined scouring by stringers of water loaded with particles. They frequently accompany flute casts. The point of initial erosion, and with it the direction of the current, are usually difficult to identify from the shape of these latter structures.

3.3.1.2 Tool Marks

These are also structures found on the base of a bed where they may result from the presence of an immobile object on the sediment surface (stationary tool mark), from the impact of a current on an object (obstacle mark), or from the impact on the sediment of an object moved by the current (moving tool

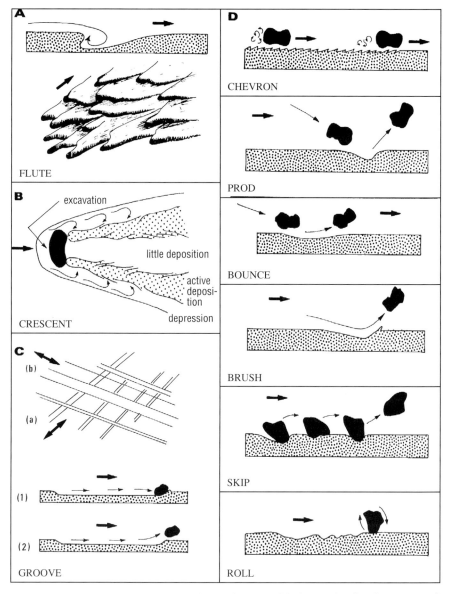

Fig. 3.13 A–D. Examples of erosion marks at the base of beds (partly after Sengupta and Ricci Lucchi, in Reineck and Singh 1980). Current from left to right. **A** Formation of flute casts, with imprint shown on the inverted bed; **B** crescent marks; **C** groove casts (non indicative of current direction): expression of casts resulting from successive currents (*a, b*); with (*1*) or without (*2*) preservation of original object at the end of the cast; **D** formation of tool marks

mark). The deposition of floating objects (e.g., tree trunks) takes place usually in zones of temporary emersion, such as flood plans and intertidal zones. Their fossilization, however, will only be possible in submerged areas subjected to active sedimentation. Obstacle marks occur mostly as crescent casts of lunate shape, the concave side pointing in the upstream direction. Resulting from the deflection of a current around an obstacle, such as pebbles or shells lying on their ventral side, these structures indicate erosion on the upstream and lateral sides of the objects, and the deposition of a sand "tail" on the downstream side. The splitting of the current around the object frequently leads to V-shaped structures enlarging in the downstream direction. The deposition of material from the water behind the object is comparable to deposition of sand behind obstacles in aeolian environments, leading to the initiation of dunes. Together with flute casts, with which they are frequently associated, crescent marks or casts are excellent markers for the younging direction of strata, and for the direction of the currents depositing these beds. Despite this, these two types of structures, together with the majority of the other sole casts, are of little use as environmental indicators. They may result from deep marine currents (flysch, base of turbidites), from storm waves, from fluvial inundations, tidal currents in sandy-clayey estuaries, etc.

Moving tool marks are highly varied in appearance due to the variety of objects, modes of transport, and the nature of the substrate. They bear witness of the orientation of the current, rarely of its direction, and virtually never of the depth of water cover. The most common ones are groove casts produced by an object sliding or rolling over an unconsolidated bottom. Sometimes the objects come to rest at the end of the grooves which may form different directional groups. Chevron casts are V-shaped, occur in a rectilinear pattern, and are closed in the downstream direction. They appear to result from objects moved in parallel above a clayey surface with periodic short turbulences on their back side. Prod casts result from the low-angle impact of objects on the bottom. They are asymmetric, triangular to half-cone-shaped, and deeper on the downstream than on the upstream side. Bounce casts result from shallower angles of impact. Being practically symmetric, they are only of general directional value. Brush casts form as the result of the very slight contact between a moving object and the bottom. Their long axes are preferentially inclined in the downstream direction. Skip, and roll casts, represent a succession of impacts from irregular objects rolled over the bottom, or transported by saltation.

3.3.2 Surface Structures or Imprints

3.3.2.1 Current Structures Under Very Thin Water Cover

In zones of temporary emersion such as alluvial plains and beaches, various characteristically delicate, but rarely fossilized, structures occur on the top side of beds: indentations produced at the sides of channels by successively lower stages of an ebb flow or a receding flood (water level marks), and irregular microripples (0.5–1 mm high, several millimeters long) resulting from the wind action on a sedimentary surface covered by only a few centimeters of water (wrinkle marks). Other structures are regular ripples on a millimeter scale forming under undirectional movement of a water film just before emersion (millimeter ripples), small sandy crests on millimeter to centimeter scale resulting from the adhesion of dry wind-transported particles on a moist sand left behind by a receding tide (antiripplets, adhesion ripples), and air bubbles trapped within beach sands during a rapid tidal flow or by moderate wave action (bubble sand structure or, when in carbonate-platform sands, bird-eyes). Also common are mm-thick laminae in the shape of large festooned arcs with the convex side pointing to the shore. They consist of relatively fine-grained sand, or easily carried particles such as micas or shells which are gradually left behind by waves during a retreat of the sea (swash marks). Marks found aligned at the high water line sometimes result from dried plant fibers. Rill marks are micro-erosional structures of a complex dendritic pattern, anastomosing in the downstream direction. They result from the erosion of sediment by small water veinlets flowing over the sand during low tide or a rapidly receding flood.

3.3.2.2 Structures Resulting from Intense Currents

When a strong current flows over a sedimentary substrate transporting rather mobile particles, particular erosion marks are formed in environments close to the emersion level. These *primary current lineations* characterize horizontal bottoms submitted to intense movements (upper plane beds) in the upper flow regime (cf. Sects. 3.2.2.2 and 3.2.2.3). The result may be *channels* scoured in the course of rivers, in the coastal zones at the outlet of bays, or in deep-sea fans by gravity flows. They may be observed on the top of beds, in their inner parts, and also at the base. Channel fill deposits frequently exhibit at their base a concentration of coarse particles and current structures. Filling may be in the form of horizontal layers, or in concave symmetrical or asymmetrical layers, depending on whether the particular channel is closer to emersion, deeply submerged, or subjected to oblique currents (Fig. 3.14). Lastly, reference has to be made to scour-and-fill structures developed asymmetrically along the axes of a channel with a steep upstream side and a shallower downstream side. They are sometimes associated with channels and impact marks (grooves, etc.), and are also known as gutter casts.

3.3.2.3 Escape and Pressure Structures

Sand or mud volcanoes are subcircular in plan, conical in section with a central crater, and vary in size from a few millimeters when they are referred to as pit-and-mound structure, to several dozens of meters. They result from the upward expulsion of a sediment-water mixture under the influence of a number of factors: gravity movement within the deposit (slumping), earthquakes, or

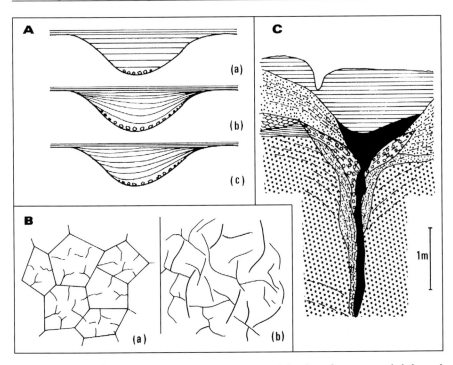

Fig. 3.14 A–C. Examples of post-depositional structures. **A** Section of current-eroded channel subsequently filled by *a* horizontal, *b* symmetrical concave, or *c* asymmetrical layers. **B** Surface expressions of shrinkage structures resulting from subaerial desiccation (*a*) and subaqueous synaeresis (*b*). **C** Ice wedge in Pleistocene sediments of England subsequently filled by heterogeneous continental sediments (after Gruhn and Bryan, in Collinson and Thompson 1982)

tectonically induced compression, as in submarine accretional prisms, subduction zones, zones of collision, rearrangement and compaction of argillaceous particles deposited in bulk, and liberation of gases derived from breakdown of organic matter, from methane in clathrates, and other endogenic sources. The sediments expelled through the crater flow in inclined layers over the periphery of the cone, thereby increasing its volume. Characteristic of regions of strong tectonic activity and irregular, but very rapid sedimentation, such as along the Caribbean, Indonesian, and South Japanese margins, sand volcanoes may be of continental as well as of deep or shallow marine origin.

Clastic dykes or sills are formed within sedimentary layers by the intrusion of exogenic unconsolidated sandy or clayey material under the influence of strong horizontal, vertical, or oblique pressure. The emplacement of the dykes and sills may take place in unconsolidated sediments, or in zones of fracturing and jointing in consolidated rocks. Their vertical extent below the surface may be several dozens of meters, and their horizontal length well up to a kilometer. They can also occur in consolidated rocks (in Beaudoin et al. 1983). The pressure required for the clastic intrusions is controlled by the sedimentary load, by gravitational movements, and by hydrostatic or gas pressure. However, often they are only the result of downward filling of an open fissure. Like sand volcanoes, clastic dykes are encountered in both continental and marine deposits.

3.3.2.4 Structures Resulting from Physico-Chemical Modifications of Exposed Sediments

A number of highly diagnostic structures found on bedding surfaces result from processes taking place after the respective beds have been exposed. Desiccation polygons or *mud cracks* formed during the subaerial shrinkage of originally water-saturated argillaceous sediments are a good example. There are, however, also structures on bedding surfaces resulting from subaqueous shrinkage, the so-called synaeresis cracks, the formation of which is controlled by vertical or horizontal tension. They may be distinguished from mud cracks by the less regular arrangement of the polygons, being partially incomplete and less wide than the former (Fig. 3.14). Also, they are not open, and not V-shaped in section. True mud cracks are made up of primary polygons within which polygons of secondary, tertiary, etc. order are developed. During burial, the fissures are filled by exogenic material which, in the case of a calcite cement, may form a network with positive relief. Associated with these structures of subaerial desiccation may be raindrop marks (circular in plan, or oval in the case of rain falling at an angle from the vertical), tracks of birds and other terrestrial animals, pseudomorphs after felted acicular ice crystals (e.g., during the phreatic escape of water on a beach during cold season), or crystals of salts and gypsum. Other structures result from deposits of foam in networks of bubbles of different sizes, patterned or without preferred orientation (foam impressions), or from ice wedges, ice cracks, or deposits filling open fissures left behind by ice wedges (Fig. 3.15).

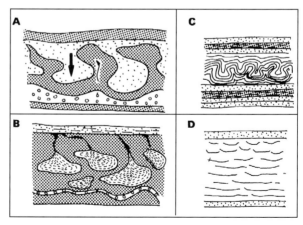

Fig. 3.15 A–D. Examples of synsedimentary mechanical deformation. **A** Load structures; **B** ball-and-pillow structures; **C** convolute bedding; **D** dish structures

3.3.3 Synsedimentary Deformation

3.3.3.1 Structures of Mechanical Origin

Within freshly deposited sediments, deformational structures can originate under the influence of differences in mechanical properties or gravity of the materials concerned. These structures can be observed mostly only in section. The degree of deformation increases as the sediment becomes saturated with water and when the degree of burial is small. These *load structures* result from the pressure exerted by sandy beds on underlying plastic clayey layers. The clay tends to infiltrate the base of the sandy beds in upward-pointed protuberances or flame structures, whereas the sand will migrate downwards in irregular, detached and deformed masses. The surface of the clayey beds exhibits load casts, as distinguished from flute casts by their irregularity and lack of preferred orientation. The formation of load casts may be favored by initial differences in mass (overloading in the sandy layer due to the presence of current ripples) or by a slight downstream inclination of the beds. Ball-and-pillow structures are found within alternations of clayey and sandy beds. They consist of loaf-shaped bodies of sand detached from the overlying sandy beds and floating in a fine-grained matrix. The sand bodies are contiguous or isolated, depending on the amount of matrix available, and tend to be more frequent in the lower portion of the beds. Experience shows that the sandy layers can become fragmented by vertical shock waves and by overloading. Horizontal forces can also be associated with this phenomenon, as is the case for a sandy sediment rapidly deposited on a low slope, and progressively covering the underlying sediment. Load casts and ball-and-pillow structures are encountered in a variety of subaqueous environments, ranging from shallow to deep water.

Intense deformation within layers of fine-grained sands, in the form of numerous uninterrupted contorted laminae, is known as convolute bedding. It is caused by processes of liquefaction within the sediment as a result of seismic shock waves, movement of phreatic water, or of currents interacting with the underlying strata along the water/sediment interface, leading to suctional effects on the beds (cf. discussion in Reineck and Singh 1980). Dish structures consist of flattened argillaceous bodies within a sandy matrix which become smaller and concave upwards as the top of the bed is approached. They result from the upward escape of pore water within a poorly permeable sediment during consolidation and burial by sand. The water is displaced laterally until it encounters vertical passageways in the clay which subsequently are filled from above by columnar sand bodies, the so-called water-escape structures. Dish structures are encountered in various environments of rapid sedimentation, such as alluvial fans, delta fronts, and the lower parts of the continental margins.

Truely gravitational sliding or slumping, a major mechanism of mass transport of sediments (Sect. 3.1.2), can also take place on a very local scale, leading to deformation without much displacement. The resulting structures are characterized by intense synsedimentary folding on different scales (contorted bedding, slump; Sect. 3.4.2). Slumping is frequently associated with fracturing and local brecciation, together with rotation and overturning of blocks of sediments, with injection phenomena, and the formation of synsedimentary faults. Slumping is encountered in various types of sediments and environments: oceanic oozes, coastal muds, dune sands, water-saturated marls on exposed slopes, etc.

3.3.3.2 Biogenic Deformation

The activity of organisms on the surface of, as well as within, sediments leads to disturbances of different intensity. This process is known as *bioturbation* and the resulting structures as trace fossils or *ichnofossils*. The study of trace fossils is a comparatively young subject, motivated by its applicability in various aspects of geology: characterization of the stratigraphic relationship of these fossils, their mode of life, faunal provinces, sedimentary gaps, depositional environments, limitations of high-resolution stratigraphic investigations, polarity of beds, structural deformation, etc. (cf. Frey 1975; Gall 1976; Frey and Pemberton, in Walker 1984). In general, the traces are better recognizable in sandy sediments with low contents of organic matter, less reworked after deposition (Fig. 3.16). Organisms disturb the sediment usually over a thickness of a few centimeters only, rarely over several dozens of centimeters and, as such, intervene during disturbances of primary depositional nature through water–sediment interaction, and early diagenetic processes.

The identification of trace fossils is based on three related principles (cf. in Reineck and Singh 1980; Collinson and Thompson 1982). The taxonomic classification is very difficult because the organisms responsible for most trace fossils are not known, and because of the fact that different organisms may produce similar structures. A classification based on the relation between the nature of the sediments involved, the exact position of a trace within a given bed, and their state of preservation (toponomy), is useful in the study of an-

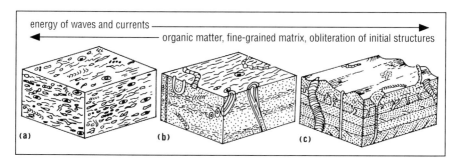

Fig. 3.16. Relation between types of bioturbation and hydrodynamic conditions of deposition (after Howard, in Collinson and Thompson 1982); *a* intensely bioturbated argillaceous silt, trace fossils compressed and little recognizable; *b* fine-grained sand, intermediate stage; *c* coarser-grained sand with preserved sedimentary structures, trace fossils usually recognizable

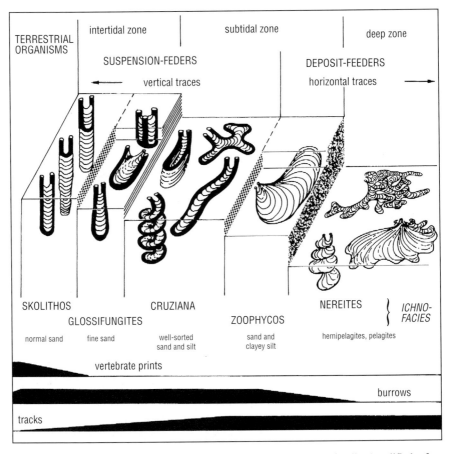

Fig. 3.17. Example of bathymetric zonation according to trace fossils (modified after Seilacher 1967 and Heckel 1972, both in Selley 1976; Reineck and Singh 1980)

cient sequences (e.g., Fig. 3.16). The most fruitful approach considers the deformation and sedimentary characteristics in relation to type of trace and the habitat of the organisms responsible for individual trace fossils. In this ecological nomenclature the following types of traces are distinguished (Seilacher 1953, in Reineck and Singh 1980): resting traces like those left by certain fish or starfish on the sea bottom, crawling traces of echinoderms, molluscs, crustaceans, etc.; browsing or grazing traces on the sediment surface from epibenthic animals like certain fish or gastropods; feeding traces of organisms within sediment layers such as various types of worms and endobenthic molluscs which digest mud or sand; dwelling traces of burrowing animals (polychetes, syphoned bivalves, etc.).

References are found in Heezen and Hollister (1971), Chamberlain (1975), and Gall (1976). Such a classification leads to a subdivision of the organisms

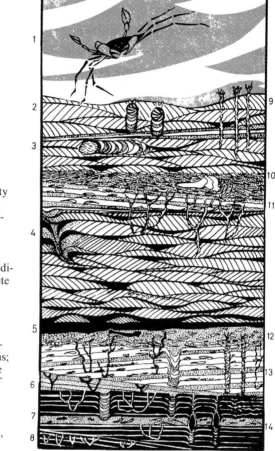

Fig. 3.18. Examples of trace fossils resulting mainly from benthic activity (after Schäfer, in Reineck and Singh 1980); *1* Crustacean on sediment surface; *2* Pelecypods moving upwards with progressive deposition of sediment; *3* Echinocardium moving laterally, thereby compressing the sediment; *4* escape structure of Polychaete worm (Aphrodite); *5* bone of bird in shelly sand covered by clay (*black*); *6* downward escape structure of Lamellibranch and worm burrows; *7* fish otolites; *8* worm burrows; *9* agglutinated tubes of Polychaete worms; *10* bone of marine mammal in coarse shelly sand; *11* dwelling structures of Polychaetes; *12* agglutinated tubes deformed in a coarse bed; *13* vertical escape structure of Gastropod; *14* lamellibranchs in situ with siphon, escape structures below

according to their depth of living, and to the correlation of the traces with environmental factors such as turbulence, currents, nutrients in suspension, etc. (Fig. 3.17). Although rather fruitful in principle, this approach leads to a number of difficulties when applying the conclusions from modern environments to ancient sediments. This is discussed in more detail by Wetzel (in Stow and Piper 1984). A number of examples of traces and disturbances left on the surface of, or within, sediments by bioturbating organisms are presented in Fig. 3.18 (from Reineck and Singh 1980).

3.4 General Mechanisms of Sedimentation and Resulting Deposits

3.4.1 Settling

Deposition by settling always takes place in areas of low hydrodynamic activity, thereby allowing the slow descent of individual particles or aggregates to the bottom. In addition to being common in lakes, the lower reaches of rivers with low slopes, and in protected shallow-marine environments such as lagoons, deltas, or sheltered bays, these deposits are particularly characteristic of open marine bottom environments. They occur usually in areas removed from the continental margins, or topographically elevated above the adjoining areas (seamounts, submarine ridges or plateaus), isolated from gravitational supply by sediment traps (such as circumpacific trenches) or distant from areas of gravitational and bottom currents. Settling deposits usually consist of planktonic tests and silty-argillaceous terrigenous particles. The sedimentation of fine-grained clays often is accelerated by their tendency to flocculate through electrochemical processes, or to agglomerate in organo-mineral complexes or fecal pellets derived from zooplankton. The main characteristics of marine settling deposits are: low to very low rate of sedimentation (in the order of 1–100 mm/1000 years); commonly continuous bioturbation with the exception of anoxic basins; compositional homogeneity within a given horizon, with the possibility of facies alternation (e.g., differences in carbonate content); absence of syndepositional or erosional sedimentary structures; and relative abundance of planktonic tests, fine-grained terrigenous particles and minerals formed in situ.

According to Berger (in Stow and Piper 1984), open-marine sediments deposited in calm waters can be subdivided into: *pelagic oozes* characterized by the preponderance of planktonic tests (above 75%); *hemipelagites* containing less than 5% tests, about 40% silt, and much terrigenous clay. Pelagic clayey oozes with 25%–75% planktonic remnants and abundant clay, as well as pelagic clays with less than 25% planktonic remnants and more than 60% clay, occupy intermediate positions. Pelagites abound in abyssal environments (Hsü and Jenkyns 1974; Stow and Piper 1984) far removed from direct terrigenous sources, mostly in zones of high organic productivity sheltered from stronger dissolution of biogenic oozes (cf. Chap. 1), and in areas receiving strong aeolian supply of material, and with prominent development of authigenic clays (cf. Chap. 4; Fig. 3.19 A). Where associated with metalliferous sediments and manganese nodules, they are often rather homogeneous and oxidized, and poor in organic matter and sedimentary structures. The hemipelagites accumulate preferentially along the continental margins, in adjacent basins, and in seas of intermediate dimensions such as the Mediterranean and the Japan Seas marked by an abundance of terrigenous supply. These sediments are characterized by the poor size sorting typical of sediments deposited fairly rapidly. They are usually made up of material from two dominant sources, one terrigenous and fine-grained, the other planktonic.

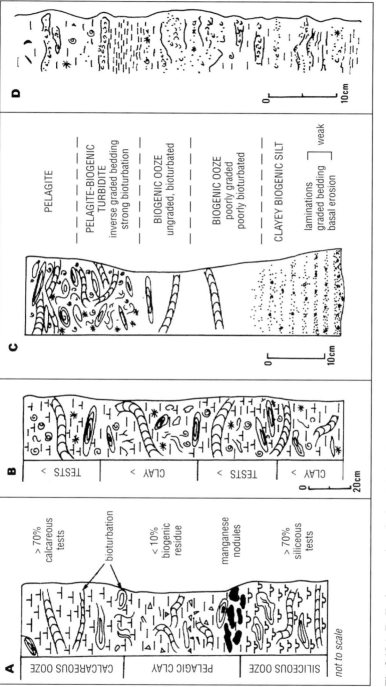

Fig. 3.19 A–D. Facies types in pelagites (**A**), hemipelagites (**B**), fine-grained turbidites (**C**), and fine-grained contourites (**D**) (after Stow and Piper 1984; see also text for further explanation)

Variations in abundance through time of these materials lead to different degrees of alternation between the two facies (Fig. 3.19 B).

Where the rhythmic sequence of sedimentary layers is pronounced, as in the alternation of limestones with marls, the term *periodites* is used. Such rhythmic deposits are observed particularly in basins with fairly rapid sedimentation accompanied by moderate bioturbation. The rhythms are frequently controlled by the alternation of pelagites rich in marine biogenic matter with hemipelagites richer in compounds of continental derivation. The causes for such alternations may be climatic in nature, especially when they are of great geographical extent, and periodic on various scales, involving changes in global parameters.

An example is the alternation of humid periods, favorable for increased argillaceous supply from the continents, with drier periods resulting in a stronger planktonic carbonate production (Einsele and Seilacher 1982). The formation of periodites may also result from various other causes such as temporary upwelling of deep waters, variations in sea-level or in the position of the carbonate compensation depth, cyclic fluctuations of the atmospheric CO_2-content required for limestone formation, etc. Certain types of alternations result from mechanisms other than pure settling, for example, subperiodic interruption of hemipelagic sedimentation through gravitational deposits derived from higher elevations, succession of tidal deposits, and diagenetic redistribution of limestone within initially comparatively homogeneous marls. For a discussion, reference is made to Einsele and Seilacher (1982).

3.4.2 Gravitational Sliding

Mass deposition under the influence of gravity, or *resedimentation*, results from a number of different processes (cf. Sect. 3.1.2). Some, such as land slides of consolidated rocks, rotational slides of unconsolidated rocks, or mud flows, occur in subaerial environments. However, the majority of resedimentation phenomena occur in subaqueous environments.

Slumps result from the displacement or slumping of a little-consolidated sedimentary mass close to the floor, without disintegration in water. They are characterized by intense deformation of beds, mainly by folding. Their dimensions can range from meters to several kilometers, and their volume can attain several cubic kilometers. When a large slump forms over a moderate slope, it will develop tension faults and large anticlinal undulations (roll-over anticlines) on the upstream side, relatively little deformed mid-sections, and tight, highly contorted folds dissected by faults on the downstream side (Fig. 3.20).

Recognition of the faults at the upper end permits a distinction as to whether the synsedimentary deformation is the result of rapid slumping or of growth faults developed slowly and gradually within an uninterrupted se-

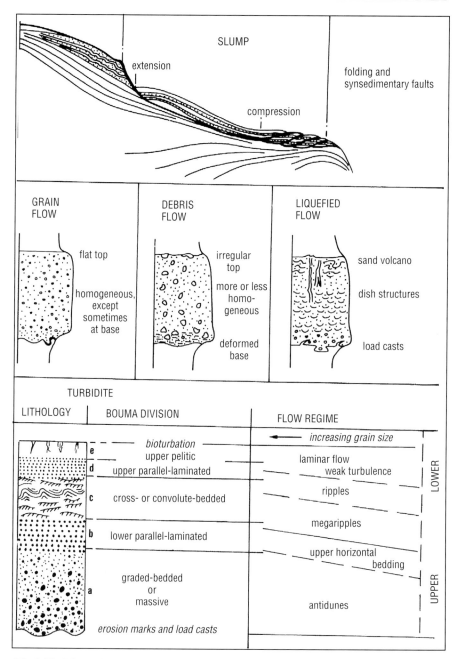

Fig. 3.20. Schematic section through large-scale slumps on moderately inclined slopes, grain flows, debris flows, and liquefied flows. Scheme of a turbidite and inferred flow regimes corresponding to Bouma divisions (in Fairbridge and Bourgeois 1978)

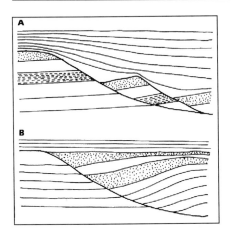

Fig. 3.21 A, B. Schematic sections through post-depositional synsedimentary faults. **A** Rapid movement produced by slumping through upstream tension; **B** slow movement compensated progressively by sedimentation (growth fault) (after Collinson and Thompson 1982)

quence (Fig. 3.21). A slump frequently is difficult to distinguish from tectonic structures along faults, or in folds of similar size, especially when its extent is large and the outcrop situation poor. A slump exhibits the following peculiarities: juxtaposition of deformed and undeformed beds, upper contacts sometimes eroded and molded by overlying beds, irregular orientation of folds due to deformation and lateral rotation during sliding; and lack of directional coincidence with the tectonic trends (Reading 1986).

The formation of slumps is favored by the following factors: earthquakes, tectonic uplift or rapid subsidence, rather steep slopes of above $4°$ average inclination, rapid deposition leading to excessive loading, sediments of contrasting nature and differing in behavior under compaction or loading (e.g., dense sands overlying less dense clays), elevated water contents, rapid biochemical degradation of organic matter resulting in the production of gas within a sediment, deformation of base due to volume expansion of evaporitic masses (halokinesis). These factors explain why slumps are particularly frequent in front of large rivers debouching into lakes or the sea, above submarine detrital fans, and along unstable edges of the continental margins, and also why they are widely present in seismically active zones (Bally 1983). They are particularly abundant and of large dimensions along the eastern coast of North America.

3.4.3 Deposition by Density or Gravity Currents

3.4.3.1 Mass Deposition

When particles transported by a gravity flow are slightly or moderately dissociated in the carrying fluid (cf. Sect. 1.2), sorting is poor and sedimentation takes place in a mass flow. These processes of resedimentation occur in three main types (Fig. 3.20).

1. Grain flows result in massive deposits with an upper plane surface, a lower zone sometimes inversely graded because of frictional effects, and scour marks. They mostly represent sand avalanches, developed especially on the back of continental dunes, or on the downstream side of subaqueous current ripples. They are known from recent and ancient depositional environments.

2. Debris flows, or debriites, represent the most frequent type of mass flows. Being a few decimeters to several meters in thickness, they are characterized by a great range in sizes from granules to boulders, floating in a silty-clayey matrix. Vertical structuring, grading, and imbrication of the clastic particles, are poorly developed.

The base of a mass flow can deeply erode an unconsolidated substrate, resulting in sedimentary structures indicative of deformation and shearing. The upper bedding surface is irregular. Debris flows are frequently encountered in submarine channels. They can deposit water-bearing, poorly consolidated conglomerates, occurring on slopes of as little as 0.1°, and locally over large areas such as in the Atlantic Ocean West of central Morocco (over 700 km length, area of 30000 km^2). The frequency of debris flows in the geological record appears to be underestimated, especially as they can be confounded with certain marine glacial deposits. Their mode of emplacement, favored by tectonic or seismic activity, is similar to that of olistostromes, thick formations sometimes containing very large blocks or olistolithes. The Alpine "wildflysch" or pebbly mudstones, deposited in submarine environments under the influence of gravity down the front of tectonic nappes, also appear to belong to this type of sedimentational phenomenon.

3. Liquefied flows, transported as a mass relatively diluted by an upward moving liquid (Fig. 3.4), result in deposits intermediate between debriites and turbidites. They are characterized by weak size grading and the presence of various types of escape structures: load casts at the base, dish structures and massive contorted beds, and sand volcanoes on the top of the beds. These deposits generally are associated with debris flows on the upslope areas of the continental rise. They are uncommon in ancient sequences, for example, the Basque flysch of northern Spain, and may correspond to certain graywacke facies.

3.4.3.2 Deposition from Currents of High Density and Velocity

The turbulent movement of a mixture of independent particles in water which is denser than the surrounding water body constitutes a *turbidity current*. Sorting is established during transport, and a horizontal and vertical organization of structures is established within the beds during deposition (Fig. 3.20). This results in a sedimentary sequence or ordered succession of lithofacies, referred to as the Bouma sequence. When completely developed, the resulting *turbidite* consists of five successive subdivisions representing stages of decreasing energy (Bouma, in Fairbridge and Bourgeois 1978):

a) massive or normal graded-bedded division with erosive structures at the base (flute and groove casts, isolated pebbles, etc.) and sometimes with inverse size grading resulting from internal friction. The lack of size grading is evidence of a well-graded sediment forming the source of the reworked material. The hydrodynamic environment is rather violent, corresponding to the energy required for the formation of antidunes.

b) The lower parallel-laminated division consists of rather coarse-grained material and gradually develops out of the underlying division. Graded-bedding is frequent, but may be masked by the laminated nature of the sediment.

c) The third division consists of foreset laminae of current ripples, deformed to varying degrees in contorted beds. The division marks the transition from the upper to the lower flow regime (cf. Sect. 3.2.2 and Fig. 3.20). When the contact with the underlying division is sharp, it marks the change from arenites to pelitic material.

d) The upper parallel-laminated division is made up of silty-clayey laminae which are more or less clearly recognizable, depending on the grain size and the intensity of postdepositional deformation. The contact with the underlying division is frequently gradual, indicating an environment of low turbulence.

e) The final, pelitic division is depleted in depositional structures, and frequently increasingly bioturbated towards the top. It is the result of slow laminar flow, usually linked to the presence of a nepheloid bottom layer of water. It grades slowly into hemipelagic and pelagic sediments, unless another turbidite is emplaced.

Complete Bouma sequences are observed comparatively seldom in the rock sequences where such structures are developed: submarine delta fronts, flanks of seamounts and submarine canyons, sedimentary environments related to zones of tectonic uplift (flysch), etc. This can be ascribed to a number of reasons. A given turbidite may be eroded to varying degrees by the following turbidite, leading to the elimination of divisions e, d, and sometimes c. Because the energy of a turbidity current tends to decrease along its path of travel, deposition of the coarser basal divisions a, b, and even c in the more distal parts of the basin, is precluded. These differences allow a distinction between proximal turbidites with Bouma divisions a–c, deposited closer to the source, from distal turbidites with divisions c–e deposited farther away in the open sea. Fluxoturbidites are proximal, coarse-grained, and poorly sorted. In this scheme which is frequently used in paleogeographic reconstructions (cf. Sect. 3.5 and Chap. 7), the complete turbidites exhibiting divisions a–e occupy an intermediate position. This topographic zonation can be complicated by peculiarities of the submarine relief and the supply of sediment. Furthermore, the energy levels and particle sizes of a turbidite issuing from sloped shallow-water environments differ from those of turbidites originating in deeper water due to bottom failure. Some investigations have shown the existence of fine-grained silty or silty-clayey turbidites made up only of division e, and consisting mainly of terrigenous or biogenic material. Such turbidites are characteristic of various bottom areas of the open ocean (Fig. 3.19 C; cf. Warme et al.

1981; Stow and Piper 1984). In contrast to this, giant or mega-turbidites can even develop over open-marine bottoms as the result of exceptional events such as earthquakes, thereby disturbing the theoretical zonation of distal and proximal turbidites. These notes underline the necessity of collecting arguments of different nature for a reconstruction as close to reality as possible: sedimentary succession, but also grain size, current directions, depth of deposition, mineralogical changes, etc.

The formation of turbidites is favored by a strong supply of sedimentary material (large river systems, biogenic accumulation on carbonate platforms, tectonic activity on the upslope side), seismic activity, by a marine bottom topography allowing the deposition of terrigenous deposits near the shelf-slope transition, by the presence of canyons draining material to the open sea, and by the majority of the factors which are also responsible for the formation of slumps. Turbidity currents cover larger areas than mass slides, especially in the basins at the foot of the continental margins where they contribute to a topographic leveling. As they are moving in the upper flow regime, they are able to erode loose or semi-consolidated underlying sediments, especially where the currents advance along submarine channels (cf. Kennett 1982; Reading 1986).

3.4.3.3 Deposition from Currents of Low Density and Velocity

As outlined above, the distal portion of classical turbidites, which is partly sorted and sometimes difficult to distinguish from hemipelagites (Fig. 3.19), results from currents of decreased velocity and density due to their prolonged travel. However, other fine-grained deposits result from turbidity currents which have been slow and of low density right from the start. These show no gradation towards the open sea, and no dampening of their energy level. This is the case for fluviatile flood supply along the coastal margins, and for the resedimentation of fine-grained sediments (coastal oozes) from the continental shelf by currents resulting from exceptional tides or storm waves.

Masses of suspended fine-grained sediment are carried to the open sea in the form of lutite-flows spreading over the slope and foot of the continental margins and adjacent basins. The resulting deposits are parallel-laminated on the millimeter and centimeter scale, and are intercalated with hemipelagites and other reworked sedimentary material. These *laminites* consist of material too fine to allow appreciable graded-bedding. They occur in periodic or non-periodic sequences, depending on whether they result from seasonal flood discharges or occasional reworking processes (Reading 1986).

3.4.4 Deposition by Bottom Currents

Gravity currents, in the classical sense, are known from the axes of submarine canyons where they control various processes of erosion, transport, and deposition. However, the existence of intense transportation and deposition parallel to the continental slope at depths of several thousands of meters, under the influence of the great thermo-haline bottom circulations, was only recognized in 1966 by C. D. Hollister, B. C. Heezen, and W. F. Ruddiman.

Detailed studies carried out along the Atlantic coast of the USA using geophysics, current measurements, submarine photography, nephelometry, and drilling, facilitated the identification of lobes and sedimentary undulations or drifts (with lengths of several hundreds of kilometers and widths of several dozens of kilometers) parallel to the base of the adjacent continental margin at average depths of 2000–4000 m. These structures mainly consist of silty-clayey material supplied from the north. Their surfaces take the shape of asymmetric ripples resulting from southwards currents, overlain by a dense turbid layer, the nepheloid bottom layer (cf. Sect. 3.1.1). These large sediment bodies are situated along the course of the bottom current of the northwestern Atlantic, the so-called western boundary undercurrent, resulting from the downward plunge of the surficial and intermediate depth waters of the Norvegian Sea (cf. Kennett 1982). This great depth current, sometimes also termed geostrophic, closely follows the bottom at variable velocities of up to 2 km/h, corresponding to about 0.5 m/s, along the oceanic topography. During its advance, it deposits sedimentary material, leading to a leveling of the morphology. C. D. Hollister proposed the term contour currents for such bottom currents, and *contourites* for the resulting sediments. The term tractionite which was proposed at about the same time, has met with little acceptance. Numerous giant sedimentary drifts made up of contourites have since been recognized, especially in the North Atlantic (Faugères et al. 1984; Stow and Holbrook, in Stow and Piper 1984), and even in the Canadian part of Lake Superior (Halfman and Johnson, in Stow and Piper 1984), and other localities.

The identification of contourites poses a number of problems, especially in ancient sequences where these deposits can be modified by diagenesis and tectonic processes. The prime problem is their distinction from turbidites which can represent facies close to those of contourites such as in distal turbidites or turbidites reworked after deposition. The earlier acceptance of these terms has led to a certain rejection of the importance of contour currents for sedimentation along the base of the continental rise (cf. Reading 1968; Stow and Lovell 1979; Stow and Piper 1984). Nevertheless, there is, indisputably, a complete transition between the mechanisms of depth currents, diluted gravity flows, and deep-sea settling, leading to complex intercalations of contourites, turbidites and hemipelagites, especially on the continental rise. According to Hollister and Heezen (in Stow and Lovell 1979) and Bouma and Hollister (in Reineck and Singh 1980), the following essential criteria facilitate the distinction of contourites from turbidites: better sorting, thinner beds (below 5 cm) and undisturbed base and top contacts, normal and inverted graded bedding, ubiquitous horizontal laminations which are in many cases outlined by heavy mineral stringers, orientation of grains parallel to the bedding, argillaceous matrix less abundant (below 5%), fossil remains rare and usually broken and worn. Recent studies have shown that if the laminations of fine-grained sand and silt, described originally from the continental margin of the US Atlantic coast, did indeed result from contour currents, they are by no means the only possible facies of contourites and thus cannot be used as the sole reference for identification of these sediments.

For Stow and Lovell (1979) and Stow and Piper (1984) the essential criteria for contourites, as opposed to turbidites and hemipelagites are:

1. irregular vertical succession of different sedimentary facies with regular or inverted graded-bedding, but absence of ordered sequences;
2. more or less continuous bioturbation, depending on rate of deposition, with preservation of current structures;
3. local or allochthonous origin of the sedimentary constituents.

Stow and Piper (1984) identified three different types of contourites:

Muddy contourites are the most frequent ones, accounting for about 75% of those in the Atlantic Ocean. They are deposited exclusively by bottom currents and are virtually unknown from ancient sequences. On detailed examination they exhibit a number of important unordered structures: varied graded-bedding, sharp or gradational contacts, rare parallel laminations, alignment of silty pockets and lenses deformed by widespread bioturbation (Fig. 3.19D). The bioclasts present are frequently broken and derived from planktonic or benthic organisms. In the latter case they are of local, deep-water origin and not derived from coastal environments like in turbidites.

The rather common *sandy-silty contourites* result from settling as well as from sedimentary reworking of bottom currents. Among others, they entail the facies described by C.D. Hollister (see above) which are used as a reference for the identification of contourites in various fossil sequences. However, they represent a considerable range of facies: irregular beds 1–20-cm thick, with sharp or gradational contacts, bioturbation, rare and isolated graded-bedding, sandy laminations and concentrations.

The *gravel-lag contourites*, consisting of sand and gravel, result from the winnowing-out of fine-grained sedimentary particles by strong bottom currents.

The rate of deposition for contourites varies from below 1 cm/1000 years to above 20 cm/1000 years, but on average is rather high, i.e., above 10 cm/1000 years. Irregular vertical sequences from muddy to gravel-lag contourites and back to fine-grained varieties (thickness 10–100 cm) have been described by Faugères et al. (1984) from the Faro sedimentary drift south of the Spanish Atlantic coast. They show the presence of current cycles, the duration (1000–30000 years) and energy levels of which have been subjected to great fluctuations with time. They appear to be related to changes of the oceanic thermo-haline circulation systems.

3.4.5 Deposition by Violent or Exceptional Currents

The daily actions of winds and fluviatile or coastal currents, of the tides and of waves and surf, leads to a variety of deposits rich in sedimentary structures. They are frequently reworked in successive cycles and subsequently only partially preserved. Within these normal deposits, which may occur in rhythmic successions, the sandy ones in particular are true reflections of the short-term evolution of the sedimentary surface environments. Other, much rarer deposits result from exceptional events which usually lead to pronounced erosion of the substrate. Inundation deposits, sometimes referred to as *inundites*, occurring on alluvial plains and in deltaic or coastal lagoons, belong to this category. Frequently graded-bedded at their base, like turbidites, they differ from the latter in the absence of fine-grained parallel laminations (division d of the Bouma sequence) below the horizon containing asymmetric current ripples, and by the presence of bioturbation structures typical of exposed environments such as root traces, subaerial burrows, palesols, etc. (Fig. 3.22).

Storm deposits or *tempestites* have been described more frequently from ancient sequences. They are generally distinguished from turbidites and inundites by the presence of multidirectional groove casts, by the absence of load casts at the bottom contacts, and the presence of hummocky cross-bedding overlain by asymmetrical wave ripples. The top portions of these

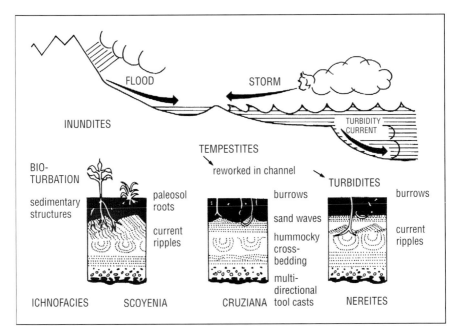

Fig. 3.22. Principal diagnostic characteristics of inundation deposits, tempestites, and turbidity currents (after Seilacher, in Einsele and Seilacher 1982)

General Mechanisms os Sedimentation and Resulting Deposits   97

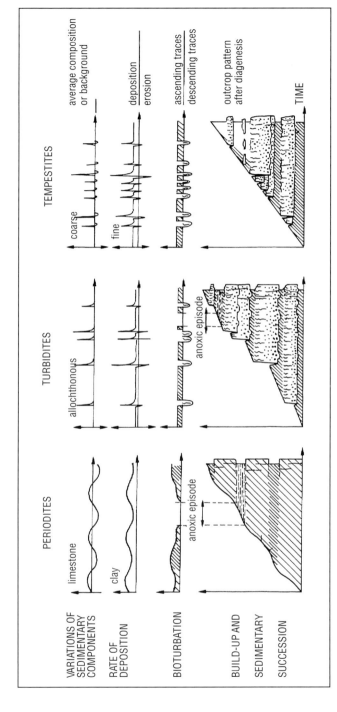

Fig. 3.23. Differences and schematic relations between periodites (e.g., marl/limestone alternations), turbidites, and tempestites (after Einsele and Seilacher 1982)

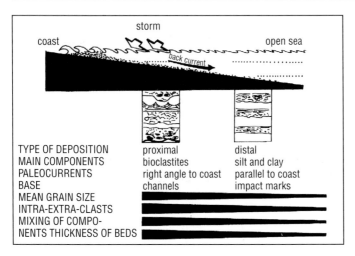

Fig. 3.24. Facies and characteristics of proximal and distal tempestites in the Muschelkalk limestone of Germany (after Aigner, in Einsele and Seilacher 1982); V lower limit of normal wave action (approx. 50 m); T lower limit of storm wave action (approx. 100 m)

deposits are bioturbated and contain fossils indicative of shallow-water environments (Seilacher, in Einsele and Seilacher 1982; Fig. 3.22). Like turbidite sequences, tempestites are marked by the alternation of periodites as a result of settling, with material brought in during sudden and violent changes in the type of material supplied and rate of sedimentation, as well as by erosional phenomena and descending bioturbation resulting from the recolonization of the bottom (Fig. 3.23).

Storm waves and tidal waves (tsunamis) may be responsible for the remobilization and accumulation of rather diverse materials: siliceous or carbonate sands, gravels, shells and other tests, bone debris, phosphates, glauconite, etc. A typical sandy tempestite comprises a sequence of deposits from normal graded-bedding, to laminations, to ripples, and in many cases eventually to a top layer of pelites, evidence of gradation from the violent upper flow regime to the lower flow regime, and then laminar flow. Such a sequence characterizes a storm current directed against the beach. This direct flow is frequently succeeded by a violent seaward return current. Its erosive potential, however, is less pronounced, the resulting beds are thinner, and the grain sizes smaller (Fig. 3.24). The detailed study of tempestites which are virtually instantaneous deposits, permits the establishment of criteria for lateral correlations, for the reconstruction of the distance to the coast and thus of the paleo-depth (cf. Aigner, in Einsele and Seilacher 1982; Aigner 1985).

3.5 Identification of Depositional Environments of Ancient Sediments: Potential and Limitations

3.5.1 Introduction

The sum of the various parameters observed in an ancient sediment facilitates the characterization of its environment of deposition, as well as the changes which this environment experienced in the course of time and space. Important factors are field and laboratory data on the sedimentary constituents, their shape and grain size and the structures and textures which the rocks acquired during sedimentation, as observed on different scales from seismic sections to hand samples. Together with the physical and biological modifications undergone by the sediments following deposition, and with the general sedimentary mechanisms, these data allow the reconstruction of the depositional environments of a given sediment and the understanding of its history. A large number of reconstructions of ancient environments, resulting from very detailed sedimentological analyses, are presented in literature. Figure 3.25 is an example in which various groups of data have been combined to ar-

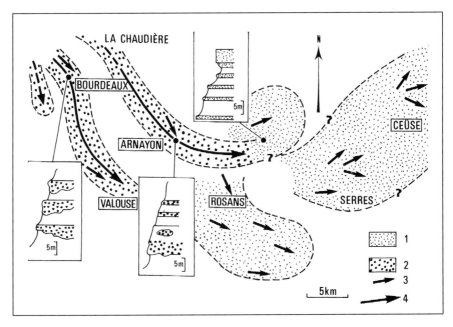

Fig. 3.25. Reconstruction of an ancient depositional environment from a sedimentological analysis (study of sequences, measurement of current directions, mapping, stratigraphic correlation): submarine detrital Ceüse fan of Gargasian age, Drôme, SE France (after Fries and Beaudoin 1984); *1* distal turbidites; *2* proximal turbidites and channels; *3* current direction as measured from orientation of fossils; *4* axes of detrital lobes.

rive at a synopsis. Such an approach facilitates a reasonable reconstruction of past conditions through the recognition of distance from the coastline, depth of deposition, current directions, and orientation of beds.

3.5.2 Distance from Coastline

In a general pattern, grain size and bed thickness of a sediment decrease in the direction of the open sea, just as land slides and slumps tend to give way

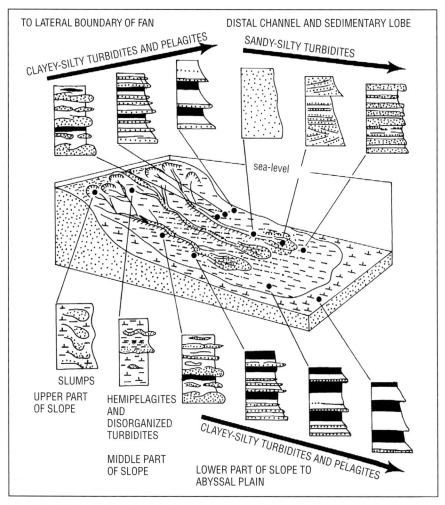

Fig. 3.26. Schematic distribution of sedimentary facies in a large detrital deep-sea fan (after Stow and Piper 1984). Note the systematic distribution of various types of fine-grained turbidites parallel and at right angles to the fan axis, as well as the presence of hemipelagites and pelagites at various levels within the fan

gradually in the downstream direction to debris flows, to proximal and then distal turbidites, to laminites and contourites, and eventually to hemipelagites and pelagites. This zonation underlies the models of deep-sea fans, the basic applicability of which has been confirmed by field experience. However, it is well known that the nature of coastal sediments of the same age and depth of deposition is mainly sandy to gravelly around capes and more silty-clayey in bays, whereas accumulations of gravel and cobbles may be encountered away from the coasts along the axes of submarine canyons within normal hemipelagites. The lateral facies variations resulting from the migration of channels of diverse depths, are classical examples of this latter situation (in Walker 1978). There is, however, a great complexity, presently being studied in detail, involving submarine morphology, types of alluviation, and mechanisms of resedimentation (cf. Sects. 3.4.2 to 3.4.5). This results in sedimentary distribution patterns which are less regularly zoned and exhibit irregular facies variations (Fig. 3.26). Great care has to be taken in such reconstructions, especially when diagenetic and tectonic reconstitutions have been intense, and outcrops sparse.

3.5.3 Depth of Deposition

Various sedimentary structures used as paleo-depth indicators, on closer scrutiny turn out to be of rather low value. This is the case for asymmetric ripples of all sizes which are encountered to great depths (Heezen and Hollister 1971), for clay pebbles sometimes found in littoral surroundings and for shrinkage structures, also developed in subaqueous environments (Sect. 3.3.2.4). Just as certain trace fossils are cautiously used as paleobathymetric indicators (Sect. 3.3.3.2), it is necessary to gather a multitude of data and to evaluate them critically. Good indications for emersion are furnished by physico-chemical modifications of air-transported material (Sect. 3.3.2.4), by sedimentary structures formed in the intertidal zone (microripples, adhesion ripples, birds-eye structures, etc.; Sect. 3.3.2.1), by imprints left by lithophage organisms such as Pholas, Cliona, worms, and barnacles, and by overflow deposits (inundites). Shallow depths are usually indicated by symmetric wave ripples, by cross-stratification, sediments of high textural maturity, and by remnants of organisms tolerating greater variations of salinity (e.g., various ostracods, bivalves, and gastropods). Other indicators are reef-building corals requiring daylight for their growth, intense bioturbation in sandy sediments depleted of fossils, or tempestites, etc. However, there are exceptions. Also, shore sediments are susceptible to becoming reworked and redeposited at much greater depths.

3.5.4 Paleocurrents

The identification of the orientation and polarity of currents is based on a number of widely applicable, clear-cut criteria (Potter and Pettijohn 1977; Miall 1984).

1. Asymmetric ripples of all sizes, cross-bedding, foreset laminae, and climbing ripples furnish unequivocal indications of the direction of the currents. However, certain fluvial and dune ripples advance in an oblique angle to the current, some wave ripples are also asymmetric, whereas antidunes generally are not.

2. Channels and scour marks are also useful indicators of orientation as they are developed in plan section, and are not too large in relation to outcrop dimensions. Seismic reference sections has proved also useful in this regard.

3. Primary current lineations, or parting lineations, formed in the upper flow regime, show the orientation of the current, but not its original direction.

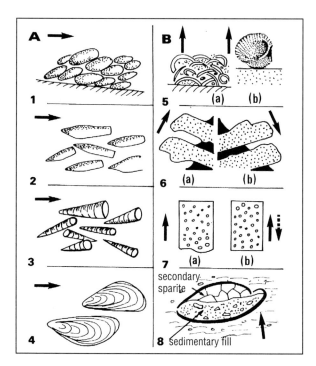

Fig. 3.27 A, B. Examples of directional sedimentary criteria. **A** Paleocurrents (*arrow* sense of flow); *1* imbricated pebbles; *2* belemnite rostra; *3* Orthoceras cones; *4* lamellibranch shells. **B** Polarity (*arrow* to top of bed); *5* bivalve accumulation by currents (*a*) or resulting from weak wave action (*b*); *6* flute casts as seen in section within a weathered sandstone, after simple uplift (*a*) or tectonic reversal (*b*); *7* normal graded-bedding (*a*) or inverted grading (*b*) resulting from progradation, lateral migration of channel, or tectonic overturning; *8* geopetal structure in Brachiopod shell

4. The upstream imbrication of pebbles in certain fluvial and littoral environments gives reliable directional information.

5. Sole casts, or erosion marks at the base of beds (flute casts, tool casts) may show either the direction of the current or only its general orientation.

6. Conical or semi-conical shells left on the surface of sediments following death of the organisms generally are deposited with the narrower end pointing in the upstream direction (belemnite rostra, Turritella, Cerithium, simple sponge spicules, asymmetric bivalve shells; Fig. 3.27). But there are exceptions: larger portion of test embedded in the sediment and turned in the upstream direction, right-hand orientation of the point of certain asymmetric gastropod shells.

7. Gradation from proximal to distal facies in turbidite or tempestite sequences, the orientation of slump folds, and lateral variations of bed thickness and grain size, constitute large-scale indicative criteria (cf. Sect. 3.4).

8. The orientation of sand grains and magnetic anisotropy are also used.

In all these cases, the reconstruction of paleocurrents assumes that the sedimentary layers containing these markers are rotated back to their original position by graphic, cartographic, or structural methods (cf. Collinson and Thompson 1982, p. 188).

3.5.5 Polarity of Beds

The identification of the younging direction of sedimentary beds, of special use in structural analyses, is based on a number of generally reliable criteria:

more or less eroded ripple crests;
cross-bedding truncating underlying beds;
normal graded-bedding, basal sedimentary structures (flute casts, tool marks);
oriented structures resulting from water escape or loading; sequences of turbidites or tempestites;
organisms buried in original living position (polyps, rudists, brachiopods, etc.);
bivalve or brachiopod shells turned by currents with their concave side facing the bottom;
buried shells partially filled with sediment, and eventually in the upper cavity filled by diagenetic spary calcite (geopetal structure, Fig. 3.27);
dwelling and escape structures of burrowing animals (Fig. 3.18), etc.

Errors may occur as with inverted graded-bedding resulting from the progradation of a delta or the lateral migration of channels, or with shells rotated by small waves onto their convex side. Problems also result from downward bioturbation in a colonized redeposited sediment or from intense postsedimentary mechanical deformation. In each case, complementary evidence must be sought and the statistical significance of the data must be considered.

4 From Sediment to Sedimentary Rock

4.1 Introduction

The term *diagenesis* covers the whole array of physical and chemical processes modifying a freshly deposited sediment, and transforming it into a solid rock. Being highly diverse in extent and nature, diagenetic phenomena depend mainly on composition and original grain size of the sediment, on the relationship along the water–sediment interface, on the porosity and particularly on the permeability of the material subjected to lithification. Another important factor is the evolution of temperature and pressure during burial or tectonic activity. Water, expelled at different rates during compaction, plays an important role as a storage or transporting medium for dissolved matter. The various mechanisms of diagenetic transformation can be identified from petrographic, mineralogical, and geochemical modifications of the original paragenetic assemblage, from chemical transformation within the cement as reconstructed from stable-isotope data, and by the localization of the sediment within stable ranges of Eh and pH. The presence of HS^-, HCO_3^-, and metal cations in the pore waters is also of indicative value, as are thermodynamic considerations. The various constituents of a sedimentary rock – carbonates, silica, clay, organic matter, etc. – are modified at quite different rates under the influence of rather diverse physico-chemical factors. Many rocks have experienced incomplete diagenetic evolution, in some cases even without notable lithification. There are very old rocks which have been subjected to only very small postdepositional modification, such as the plastic Leningrad clays of Precambrian age.

We shall limit ourselves here to a simple introduction to diagenesis, referring the reader for more detail to special treatises of the respective geochemical mechanisms (Krauskopf 1967; Steinberg et al. 1978, 1979), and the phenomenon of diagenesis in the strict sense (Berner 1971; Larsen and Chilingar 1979, 1983). We shall discuss here the diagenetic evolution of sandstones, shales, and carbonates, of silica and of fossil organic rocks.

Diagenetic evolution can be subdivided schematically into four principal stages of quite different duration, during which various phenomena make their appearance, reach a peak development, and then gradually fade away (Fig. 4.1):

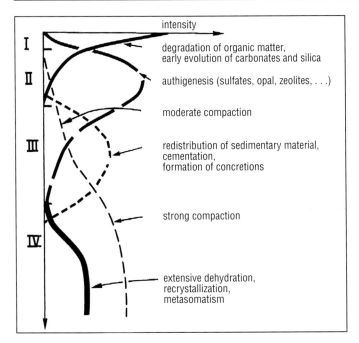

Fig. 4.1. Main diagenetic phenomena

a) *Biochemical diagenesis* is supported by microbial activity, and is closely associated with the early development of carbonates, silica, and certain evaporites. It is designated as Stage I. The majority of the organic matter, especially that of faunal origin, is destroyed, and the aragonitic tests, the fragile calcitic tests, and a large proportion of the biogenic silica are dissolved (cf. Chap. 1). At the same time, a number of gaseous decomposition products (CO_2, H_2O, NH_3) are formed. The term halmyrolysis covers the prediagenetic processes of alteration and crystallization taking place at the water – sediment interface.

b) *Authigenic formation* of sulfides, metal oxides, and of silicates (opal, zeolites, clay) takes place mainly during Stage II, when interstitial waters charged with the products of dissolution of organic matter are still able to circulate freely within the sediment prior to the onset of compaction.

c) *Cementation* takes place essentially during Stage III. It leads to filling of the voids within the rock by precipitation over long distances. Amongst the products of precipitation are opal, quartz, calcite, dolomite, iron oxides, clay, etc.

d) Stage III is also marked by the *formation of concretions*. These structures result from the local concentration of chemical compounds mobilized in the preceding stages, and their precipitation around a nucleus. Nodules and concretions consist of marcasite, flint, carbonates, sulfates, and various types of phosphates.

e) Intense *dehydration* starts during Stage IV and is accompanied by various phenomena of *recrystallization,* especially around points of grain contacts where preferential solution under pressure takes place. This leads to secondary porosity, stylolitic sutures, cone-in-cone structures, etc. The rock becomes more homogeneous due to the filling of the residual voids by authigenic minerals, the chemical constituents of which are usually of local derivation. Such minerals are salts, silica, quartz, feldspars, and clay.

f) *Compaction* also becomes pronounced during Stage IV. It commences during early burial and leads to a certain reorientation of particles, as in shales. It results also in load structures, the expulsion of water from unconsolidated sediment, and to shrinkage (cf. Sects. 3.3.2.3 and 3.3.3.1). At greater depths, it leads to flattening of fossils, mutual indentation of grains, and to complex fracturing.

g) Metasomatism or *epigenesis* concerns the replacement of certain minerals by other, more stable ones under great pressures and sometimes elevated temperatures, without change of morphology. Such processes include slow dolomitization and silicification, certain ferruginization phenomena, depolymerization, and loss of volatile matter from fossil fuels, etc.

Beyond Stage IV, diagenesis under increasing pressure and temperature passes through intermediate processes of anchimetamorphism into various stages of metamorphism and eventually into anatexis.

4.2 Formation of Sandstones

4.2.1 Main Stages and Classification

The transformation of porous sands into compact sandstones occurs through a number of successive stages largely controlled by the thermodynamic gradient during burial (Dapples, in Larsen and Chilingar 1979):

a) During the "redomorphic" stage, redox processes predominate and generally determine the final color of the resulting sandstones: oxidation of organic matter and formation of red ferric oxide coatings in porous sands; reduction and preservation of gray to black organic matter together with formation of sulfides in clayey sands or sands from stagnant-water environments.

b) During the "locomorphic" stage, cementation takes place mainly by carbonates or silica.

c) The "phyllomorphic" stage corresponds to the development of clay minerals and micas, and in particular characterizes late diagenesis and the passage into metamorphism.

These three stages do not necessarily follow each other with increasing depth. In part, they may occur even concurrently. Clay minerals can already form

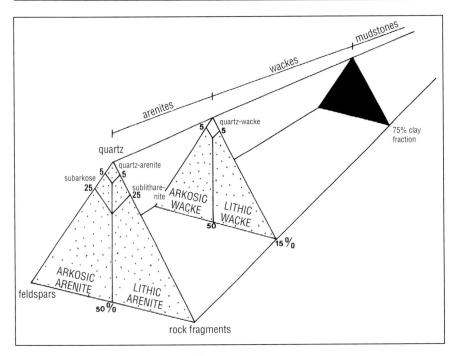

Fig. 4.2. Classification of arenaceous sediments

under moderate burial and secondary porosity can develop in already cemented rocks. Furthermore, a sandstone is commonly subjected to modifications during its diagenetic history by circulation of subterranean waters and by tectonic uplift. These evolutionary stages, which are independant of the previous ones, are referred to as epidiagenesis. It is thus preferable to examine the diagenesis of sandstones on the basis of the important mechanisms controlled by porosity and circulation of fluids. The phenomena of dissolution and cementation modify the general appearance and the characteristics of the original sediment to such an extent, that a special classification for consolidated rocks has been established (Fig. 4.2). This classification is based on the relative proportions of quartz, feldspar, and rock fragments, in combination with the amount of matrix, the intergranular phase of more or less diagenetic origin. Rocks with more than 15% matrix are called wackes, those with more than 75% mudstones.

4.2.2 Surface Phenomena

Sandy sediments deposited in arid climates (desert or alluvial plains, episodic sheet flood deposits) are immature and highly porous and, prior to becoming buried, may remain exposed over long periods in oxygenated surface environments and vadose regimes. Three types of early modifications can occur under these conditions:

a) The infiltration of detrital clays between sand grains takes place after coverage of the sediment during alluvial sheet floods by waters charged with suspended matter. This results in a clay film around the grains and gives to such rocks the appearance of matrix-rich mud flows (cf. Sect. 3.4.3.1).

b) The hydrolytic dissolution of less resistant minerals, such as pyroxenes, amphiboles, and plagioclase, is witnessed by corrosive embayments and fissures. These cavities are frequently lined with secondary allochthonous clays, protecting the mineral against further solution by infiltrating water. Sometimes the dissolved portions of the sand-sized grains are replaced along cleavage planes and within the grains themselves by authigenic minerals utilizing for their own growth the cations liberated by the dissolution, especially K, Ca, Mg, Fe, and Si. Such minerals are K-feldspars, zeolites, quartz, mixed-layer illite-smectite clays, hematite, and calcite.

c) The iron extracted by the vadose waters from black micas, amphiboles, pyroxenes, olivines, etc. contributes to the red coatings characterizing continental sandstones of arid climates such as the Permo-Triassic of Europe (Millot, Perriaux, and Lucas, in Millot 1964). The diagenetic origin of this ferruginous pigment is proven by its absence from the contact areas between adjacent grains. The pigment may be mobilized also in weathering profiles in well-drained environments to form laterites. Kaolinite is frequently associated with such corroded grains and coatings. It should be pointed out that the red coloration can be caused already by concentrations of ferric oxides as low as 0.1%.

Dune sands sometimes also show alterations as described above: clay coatings resulting from airborne dust and in places forming the matrix, coatings by ferric oxides, partly dissolved grains overgrown by authigenic minerals. Among the latter phenomena, reference should be made to the syntaxial growth of quartz, sometimes with bipyramidal terminations, on pre-existing quartz grains, the respective new material probably having been derived through aeolian corrosion. In such sections or the adjoining sands, a subterranean water layer may develop, within which evaporitic cementation by halite or gypsum takes place. Early incomplete cementation can protect the rock from later, more complete cementation. Subsequent encrustation by silica, leading to the formation of silcretes, is set between pedogenesis and diagenesis. Thus is determined the storage potential for hydrocarbons in sandstones modified by surface diagenesis.

4.2.3 Phenomena of Deeper Levels

The evolution of little consolidated sandstones depends on processes of weathering, dissolution, and compaction of the original grains, as well as on the supply of the dissolved matter required for cementation. The necessary solutions sometimes have traveled over large distances due to the elevated porosity and permeability of the rocks. The dissolution affects mainly the K-feldspars and volcanic rock fragments, thus furnishing the elements required for the formation of clays. Such authigenic clays are commonly encountered in sandstones (cf. Millot 1970; Larsen and Chilingar 1979, 1983). Smectites, chlorites, and mixed-layer minerals can form shortly after burial in the voids of basic volcaniclastic rocks. When the pore waters are acid and mobile within the sandstone body, or when the sandstone is tectonically uplifted and comes into contact with meteoric waters, mostly kaolinite will form authigenically in the characteristic form of stacked large hexagonal plates. From a burial depth of about 2.5 km onwards, random mixed-layer minerals and smectites are progressively transformed into illite or chlorite, and kaolinite tends to become replaced by chlorite (cf. also Sect. 4.3.2). The successive stages of intergranular growth of clay will not develop when the sandstone has not been pervaded by aggressive fluids. In this case, the more impermeable argillaceous intercalations will show the same clay mineralogy as the sandstones. Furthermore, the growth of clay minerals can be interrupted at intermediate stages by the invasion of hydrocarbons (in Chamley 1989). When the secondary clayey matrix becomes abundant, the original nature of the sandstone is modified and the rock develops into a wacke (cf. Fig. 4.2). This diagenetic evolution, which is characterized by the preservation of unaltered "islands" within the rock, together with reworked sediment from more distal turbidites, is one of the possible origins of graywackes, and other dark-colored matrix-rich sandstones.

The widespread siliceous cement is derived mainly from interstitial silica liberated by the destruction of K-feldspars, by the transformation of smectite into illite, and, to a lesser extent, from the solution of quartz under high pressure. This cement grows around the original quartz grains in successive layers difficult to identify in the absence of impurities, even with optical techniques such as cathodoluminescence. The complete cementation of a sandstone by silica may require up to 200 million years. Carbonate cement in sandstones in the majority of cases is calcitic, but sometimes also consists of dolomite or ferriferous calcite, ankerite, or siderite. It usually is the result of late-diagenetic migration of ion-charged solutions, notably from marly to clayey interlayers within sandstone sequences. However, in the littoral environment of hot climates an early cementation by calcite commonly occurs at low tides due to evaporation of waters rich in dissolved carbonates. These are the so-called beach rocks (cf. Sect. 4.2.2.2; Scholle and Spearing 1982).

A secondary porosity can develop within sandstones through processes such as fracturing, shrinkage, dissolution of individual particles or of the cement, etc. This acquisition of a secondary porosity can occur during various stages in the diagenetic evolution, but appears to be characteristic of particular zones of advanced depth at which dissolution of carbonates and decarboxylation take place (Fig. 4.3). The solution of carbonate cement is supported by brines containing organic acids. The ions removed, notably Mg^{2+},

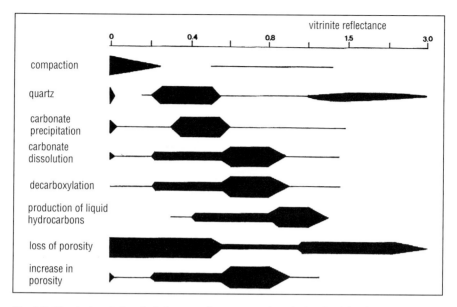

Fig. 4.3. Physical and chemical changes of quartz-arenites during burial (after Schmidt and McDonald, in Scholle and Schulger 1979). The increasing vitrinite reflectance corresponds to an increase in the degree of diagenesis.

Ca^{2+}, HCO_3^-, CO_3^{2-}, may be reprecipitated as carbonates in open pores at higher levels within the sandstones. The same pores are also susceptible to the accumulation of hydrocarbons migrating from their source rocks (cf. Sect. 4.6).

4.3 Evolution of Clays

4.3.1 Early Diagenesis

The low permeability of clays tends to resist notable fluid exchanges. Thus, after burial, clays frequently retain the characteristics of the continental or sedimentary environments in which they formed (Millot 1970; Dunoyer, in Fairbridge and Bourgeois 1978; Singer and Müller, in Larsen and Chilingar 1983; Chamley 1989). This explains why especially argillaceous sequences are frequently utilized for the reconstruction of ancient environments in sedimentary sequences sometimes measuring several thousands of meters (e.g., Fig. 1.5). Notable modifications of argillaceous successions will usually not take place during early diagenesis, unless sedimentation is very slow, their sedimentary constituents are particularly unstable, or sometimes the salinity of the marine or interstitial solutions is high.

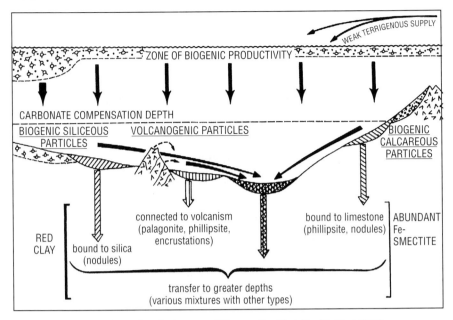

Fig. 4.4. Four types of red clays in the central Eastern Pacific (after Hoffert 1980)

a) The formation of smectites occurs at great depths in the central Eastern Pacific Ocean. There, prolonged exchanges between sea-water and the red abyssal oozes, in combination with minimal continental supply, lead to extremely low rates of sedimentation, in the range of less than 1 mm/1000 years (Fig. 4.4). The oozes form under conditions intermediate between halmyrolysis and diagenesis. The formation of clayey granules rich in iron (glauconite, verdine, pseudo-berthierine) also takes place in relatively confined microenvironments; examples are individual tests or altered minerals, close to the surface of slowly accumulating sediments along continental margins covered by temperate to warm waters, generally less than 400 m deep (Odin and Matter 1981).

b) The alteration of volcanic glass can also lead to the rapid formation of clays dominated by ferriferous smectites, sometimes associated with zeolites (especially philippsite, analcite, heulandite-clinoptilolite) with celadonite, and also palygorskite (Singer and Müller, in Larsen and Chilingar 1983). Bentonites (subaqueously deposited clayey beds frequently enriched in smectites) and the kaolinitic tonsteins associated with coal beds, also result in the majority of cases from the early alteration of volcanic glass. For an in-depth discussion, reference is made to Fisher and Schmincke (1984).

c) During deposition or early diagenesis, strong concentration of solutions in marine or continental brines, in confined alkaline basins under warm climates also can lead to the widespread formation of clays with evaporitic affinities such as palygorskite, sepiolite, and magnesian smectites. This situation was observed in the peri-marine basins of the Atlantic and the western Tethys during the lower Tertiary. In hypersaline environments like the European Triassic, corrensite followed by diagenetic chlorite may have formed (Lucas, in Millot 1964; cf. also Singer and Müller, in Larsen and Chilingar 1983).

d) The various clays formed near the surface of sediments are susceptible to becoming reworked to the detrital stage. Thus, as a result of bottom currents, red abyssal clays and abyssal oozes frequently accumulate in depressions (Fig. 4.4) or result from direct reworking over more elevated portions of the ocean bottom. Fibrous clays, on the other hand, easily fall victim to aeolian or marine transport (cf. Chamley 1989).

In normal marine environments where the rate of sedimentation is above 1 cm/1000 years and which do not exhibit the peculiarities described above, early diagenetic modifications of argillaceous sediments are frequently limited to an exchange of certain ions along the surface and in the interfoliar spaces of the clay minerals. Especially the Ca^{2+} fixed by detrital clays is exchanged against Mg^{2+}, K^+, and Na^+ from sea-water, resulting in a complete reorganization of the crystal lattice. As diagenesis gets under way in the sediment, the interstitial environment can imprint itself by morphological reorganization of the clay particles into an ordered lattice (e.g., lathed shape) and by migration of chemical compounds on a very local scale (e.g., Holtzapffel et al. 1985). Note also that the degradation of clay occurs in certain reducing sediments such as the Quaternary sapropelites of the eastern Mediterranean where organic acids lead to the selective alteration of minerals. Palygorskites here are the most easily altered minerals, and kaolinites the most resistant ones (e.g., in Schlanger and Cita 1982).

4.3.2 Late Diagenesis

The compaction of argillaceous sediments leads to textural reorganization, namely the parallel texture of shales, to a loss of porosity, and to a growing adsorption of bivalent ions. It takes place rapidly during the early stages of burial and continues at progressively slower rates at greater depth, leading to a transformation of the original clay into a shale, and eventually into a schist when tectonic processes take over. This alteration is accompanied by an evolution of the organic matter. The recrystallization of the phyllosilicates is practically complete at a depth of about 10 000 m at temperatures of around 200° C, under normal geothermal gradients. The transformation of clay minerals during burial diagenesis takes place in various stages, depending on the duration and the original minerals present (Fig. 4.5), and results in a uniform illitic and chloritic composition (Dunoyer, in Fairbridge and Bourgeois 1978; Singer and Müller, in Larsen and Chilingar 1983).

a) Smectites and even vermiculites are transformed into illites through incorporation of potassium liberated by the decay of K-feldspars, whereas chlorites use Mg or Fe from the interstitial solutions. This gradual dis-

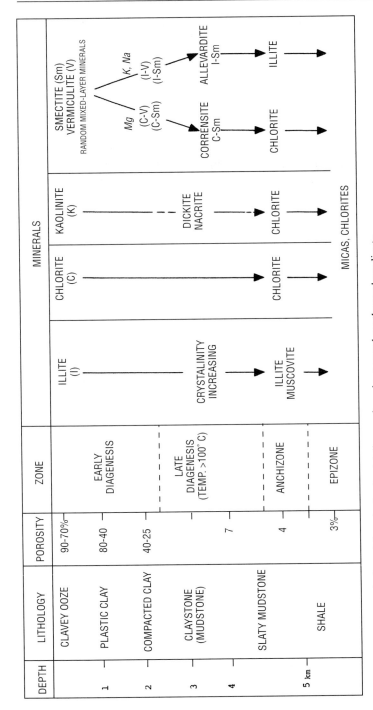

Fig. 4.5. Schematic evolution of clays during burial diagenesis under normal geothermal gradients

appearance of smectites accelerates from 2000 m onwards and is completed at around 3000 m. It frequently is accompanied by a transition of the mixed-layer minerals from an unordered to an ordered state, either sodic-potassic (e.g., allevardite = illite-smectite) or magnesian (e.g., corrensite = chlorite-smectite) in composition. In certain confined calcic or magnesian carbonate environments, highly ordered smectites and corrensite may adapt themselves to the conditions of deep-burial diagenesis or epimetamorphism (cf. Bouquillon et al. 1984).

b) Kaolinite disappears at somewhat deeper levels than smectite, leading to chlorite and even illite. Dickite and nacrite can crystallize in acidic environments, where they constitute polytypes more stable at those temperatures than kaolinite.

c) During burial, illites develop into increasingly ordered modifications, leading to a loss of crystallinity, the extent of which can be used to characterize the passage from deep diagenesis to metamorphism.

d) Trioctahedral chlorites (cf. Sect. 1.1.3) grow notably in the deep-level stages of diagenesis, whereas the dioctahedral ones suffer only minor modifications.

The geochemical evolution of clays during burial is characterized by interfoliar dehydration, adsorption of K, Na, and Mg, by the growth of the crystals, and by an increased ordering of the lattice due to ionic redistribution. In general, as one passes from smectites on surface to muscovites at depth, the 2:1 clays show an increase of the tetrahedral and of the total charge supporting the fixation of the alkali elements K and Na (Fig. 4.6). However, the in-

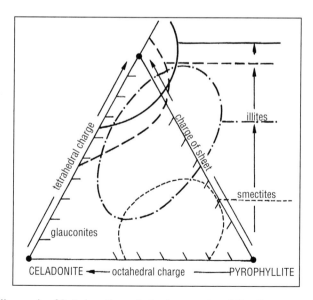

Fig. 4.6. Expression of late diagenesis of 2:1 clays through sheet parameters (after Dunoyer, in Fairbridge and Bourgeois 1978). Note: charge of sheet = octahedral + tetrahedral charge

tensity of diagenesis varies not only with the depth of burial, but also with the geothermal gradient, with tectonic activity (diagenesis and metamorphism as the result of lateral compression), and the original nature of the rocks concerned (cf. Singer and Müller, in Larsen and Chilingar 1983). The influence of the lithology on diagenesis is particularly pronounced in limestone-marl alternations (cf. Einsele and Seilacher 1982) where the limestone layers may show an enrichment of ferriferous chlorites (Deconinck and Debrabant 1985). Illites appear to be more abundant in ancient sediments than in recent ones (cf. Blatt 1982). This, however, cannot be considered the rule, and various Paleozoic and even Precambrian rocks are devoid of notable signs of argillaceous diagenesis, especially in the little deformed series of the old African cratonic basins. Thus the diagenetic modifications are only one of the possible causes for the relative abundance of illite in pre-Mesozoic sequences, other being peculiarities in climate, surface weathering of rocks, and volcanic activity (in Blatt 1982).

4.4 Carbonate Diagenesis

4.4.1 Introduction

The development of sedimentary carbonates is complex for a number of reasons. Diagenesis follows sedimentation almost immediately and the two sometimes differ only little from each other in appearance, as shown by the similarity between algal micritization of grains and certain early cementing processes. The carbonates also are easily dissolved and precipitated at various stages of their history, and their main constituents, calcite and aragonite, occupy different fields of stability, depending on environment and depth (Berner 1971). Influences by organic activity are controlled by changes in the CO_2-content, whereas the precipitation of dolomite depends more on the ionic concentrations, etc. These peculiarities explain the rather complex and essentially descriptive nature of the classifications applied (cf. Table 1.5; Fig. 1.6), and the great number of modern technical methods employed in the study of these rocks: thin sections for microfacies, selective staining, isotope geochemistry, electron microscopy, microchemistry, and cathodoluminescence, etc. As the detailed study of the various types of cement and recrystallized phases (Fig. 4.7) is highly useful in the identification of the hydrocarbon reservoir potential of carbonates, it has resulted in a wide knowledge of the processes of carbonate diagenesis (cf. Chilingar, Bissel and Wolf, also Cook and Egbert, both in Larsen and Chilingar 1979, 1983; Purser 1980, 1983; Zenger et al. 1980). This permits the identification of the successive stages in the development of these sediments.

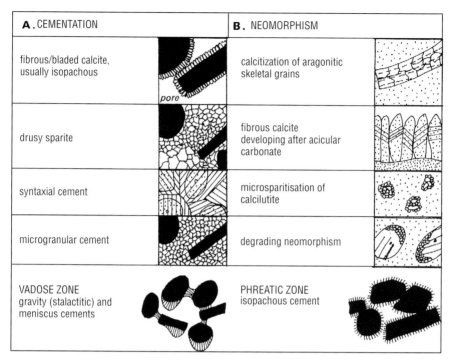

Fig. 4.7 A, B. Main types of carbonate cementation (**A**) and recrystallizations (**B**) as seen under the optical microscope (after Tucker 1982)

4.4.2 Calcitic Cement

4.4.2.1 Continental Environment

The cementation of marine carbonates sets in as a consequence of a eustatic drop in sea-level, seaward progradation of a sedimentary wedge, or of tectonic uplift. The circulation of waters in an undersaturated environment leads to the formation of sparry cement with crystal sizes increasing in the direction of the centers of the voids (drusy calcite; Fig. 4.7; Table 4.1). In the surficial vadose zone, where water occupies the pores only temporarily, the cement resulting from vertical percolation is irregular and asymmetric (Fig. 4.7), consisting of small crystals which began growing from a large number of nuclei. It is relatively rich in strontium. Four different phases have been distinguished in freshwater cements of carbonates on Bermuda (Land, in Purser 1980; Leeder 1982):

a) precipitation of low-Mg calcite on pre-existing grains in drusy or syntaxial pattern (e.g., on echinoderm debris, Fig. 4.7);

b) loss of Mg from grains of magnesian calcite without structural modifications;

c) dissolution of aragonite grains, disappearance of organogenic structures, temporary increase in porosity, start of calcite precipitation;

d) finalization of sparry cement formation, partial recrystallization; achievement of about 20% porosity, characteristic of early diagenesis.

In deeper phreatic zones, in which water pervades the complete sediment and moves laterally, regular and fairly rapid cementation leads to large crystals of uniform appearance (isopachous, Fig. 4.7), impoverished in Sr due to a leaching effect. The transformation of marine aragonite into continental sparry cement takes place through an intermediate, friable chalky zone around the grain periphery. The various truly continental cementation processes of calcite approach in appearance the pedogenic mechanisms described from such environments: pedogenic crusts on rocks containing calcite (caliche) or devoid of this mineral (calcrete), stalagmites and stalactites, pisolites etc. of karst regions, tufa and travertines resulting from precipitation.

4.4.2.2 Marine Environment

Textural, mineralogical, and chemical characteristics of the cement allow a fairly narrow definition of facies (Table 4.1; Fig. 4.8). In the *intertidal zone*, the highly varied types of cementation are mostly marked by the development of beach rocks which are common in tropical and subtropical regions, and locally present also in temperate climates. Beach rocks form under very shallow water cover of only a few dozens of centimeters through cementation by acicular aragonite or Mg-rich micritic calcite. This cementation is accompanied by a volume increase, leading to secondary deformation such as localized uplift in tepee structures. The occasional presence of asymmetric cement, also as fenestrae or birds-eyes in cavities resulting from bubbles of air or gas, as well as the destruction of algal laminae and the presence of shrinkage structures, show that beach rocks can also form in the vadose zone. The formation of beach rocks can be ascribed to a number of mechanisms. This is most commonly cementation by evaporation of sea-water and loss of CO_2 during falling tides, and by biochemical precipitation through decomposition of organic matter derived mainly from algae. Another possible process is physico-chemical precipitation in the zone of mixing between fresh- and sea-water under possible liberation of CO_2 of organic derivation (cf. discussion in Purser 1980; Leeder 1982).

In the *subtidal zone*, aragonite and Mg-calcite cements lead to the formation of aggregates called grapestones (cf. Sect. 1.2.5.3), to hard-grounds, and to the consolidation of reef complexes. Modern carbonate hard-grounds form almost exclusively over shoals. However, in the Persian Gulf off Qatar, lithification takes place on the shelf between 20–40 m depth where moderately agitated sea-water is oversaturated in carbonates and where the sands are transported very little (Purser 1980).

In *deeper marine waters,* micritic carbonate cement is formed especially when sedimentation rates are very low and the water is relatively warm and

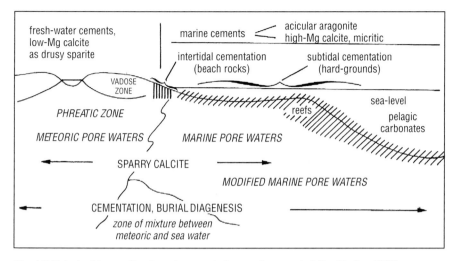

Fig. 4.8. Principal types of carbonate cementation environments (after Tucker 1982)

Table 4.1. Main characteristics of calcareous cements in continental and coastal marine environments

	Continental environment	Marine and coastal environments
Fabric	Drusy	Acicular, fibrous or platy
Minerals	Low-Mg calcite with little Sr	Aragonite, high-Mg calcite rich in Sr (up to 10 000 ppm)
Mg-Ca ratios	Low (fresh water: 0.5)	High (sea water: 3) 14–19% $MgCO_3$

saline, as in the trenches of the eastern Mediterranean or in the Red Sea. Calcareous crust and nodules of high-Mg calcite, aragonite, and even protodolomite can result from the cementation of different types of sedimentary debris down to water depths of 3500 m. They exhibit pronounced local variations in size, thickness, fabric, and chemical composition (cf. Milliman 1974; Müller and Fabricius, in Hsü and Jenkyns 1984).

4.4.2.3 Burial Diagenesis

The carbonate oozes are transformed into chalk and then into hard limestone from various depths below the sediment surface onwards, in the oceanic environment frequently exceeding 500 m (Garrison, in Warme et al. 1981). The calcitic cementation of the larger particles takes place through the dissolution of the smallest calcite particles. The porosity of Cenozoic pelagic limestones encountered during deep-sea drilling can reach 40%.

During late diagenesis of limestones when porosity can decrease to 5%, other as yet poorly understood steps of cementation may follow. The first generation of cement frequently consists of a thin coating of fibrous calcite, and even micrite in the case of grainstones, lining the cavities between individual grains. The second generation essentially occupies the voids left by early diagenesis. It consists of clear drusy sparite poor in inclusions, with plane grain contacts and frequently oriented normal to the stratification. As a result of these various modifications (cf. Larsen and Chilingar 1979, 1983), Cenozoic and Mesozoic limestones are able to preserve at least partial evidence of the syndepositional environment by way of their geochemical characteristics (Sr, Mg, oxygen and carbon isotopes). This facilitates a "chemostratigraphic" approach in the study of pelagic limestones (Renard 1986).

4.4.2.4 Additional Mechanisms

The recrystallization or authigenic formation of aragonite in calcite, or of one type of calcite in another one, can take place in a number of different ways (Fig. 4.7 B): calcitic replacement of aragonitic skelets with dark organic relics tracing more or less clearly the original internal structure of the tests; fibrous calcitic replacement of acicular calcite cement (witnessed by undulatory extinction or impurities in the fibers); microsparitization by particles 4–10 µm in size or pseudosparitization (10–50 µm) of calcilutites in irregular-shaped dispersed areas, in which the crystals have irregular or curved contacts; degradational authigenesis, i.e., the replacement of large crystals by small calcite crystals, which is a rarer phenomenon. Diagenetic stylolites usually form in limestone horizons rich in clay, parallel to the bedding or along the contact between limestone and clay beds or around calcareous nodules. They are characterized by rough plane surfaces, which in detail are highly irregular and covered by a thin lining of insoluble residue such as clay, organic matter, or metallic oxides. They form as a result of the dissolution of limestone under pressure at early or late stages, when up to 40% of the original limestone is dissolved, thereby becoming available for various processes of cementation and recrystallization. The resulting stylolites are sometimes overprinted by stylolites resulting from tectonic pressures. Cone-in-cone structures within marly limestones appear to be induced by stacked conical fractures during early diagenesis under pressure resulting from the transformation of aragonite into fibrous calcite. Lastly, note the frequent presence in limestones of diagenetic nodules of silica, phosphates, baryte, etc. resulting from the early diagenetic precipitation in deformed sediment layers, or, at a late stage in rigid undeformed sediments. They form around a nucleus and obtain their chemical constituents (Ca, Si, Fe, S, P, Ba, etc.) from pore waters oversaturated in these elements.

4.4.3 Conditions for Dolomitization

The formation of dolomite entails the growth of a crystal lattice in which calcic and magnesian layers are arranged in a regular pattern. It can take place only through very slow crystallization from a dilute solution or, more rapidly, from a concentrated solution with an elevated Mg-Ca ratio (Berner 1971). Sea-water is undersaturated against dolomite and thus almost never fulfills the conditions required for direct precipitation of this mineral. The precipitation of dolomite essentially takes place in an indirect manner controlled by exogenic geodynamic factors. During early diagenesis, there is usually a secondary precipitation of little-ordered protodolomite which has been described locally from certain littoral environments (Purser 1980). It also takes place along continental margins which exhibit relatively low sedimentation rates of below 50 cm/1000 years, and at the same time are rich in organic matter (above 0.5% C_{org}) (Baker and Burns 1985). During later diagenesis, dolomitization is virtually independant of the lithofacies, and even cuts across sedimentary units, being massive or more localized. It can take place along sedimentary disconformities and is sometimes controlled by tectonic structures. There are several models of dolomitization (Chilingar, Zenger, Bissell and Wolf, in Larsen and Chilingar 1979; Zenger et al. 1980):

a) The "infiltration-reflux" model postulates the migration of sea-water in intertidal sediments under the influence of tidal flows and capillary action, a relative concentration of Mg through evaporation, and finally dolomitization during the reflux of the waters during low tide (Fig. 4.9 A). It has so far not been possible to find examples for this process in recent sedimentary environments.

b) The "evaporative pumping" model is also applicable to littoral environments and has been proposed for the sabkhas of the Persian Gulf. It requires a continuous supply of water through the sediment, followed by intense evaporation in the coastal lagoons. The precipitation of aragonite and gypsum results in the relative enrichment of Mg in the brines. During the further passage of the brines, this element becomes precipitated and leads to the dolomitization of the host sediment (Fig. 4.9 B).

c) In submerged deposits, such as reefs or infratidal carbonates, dolomitization appears to result from slow precipitation out of weak interstitial solutions permeating the limestone. Theoretical calculations show that a mixture of freshwater with 5%–30% sea-water is undersaturated against calcite and oversaturated against dolomite (Badiozamani, in Leeder 1982). In such a mixture, the Mg-Ca ratio remains high and close to that of the initial sea-water, leading to conditions favorable for the precipitation of dolomite (Folk and Land 1975). This model of mixing two interstitial solutions of different chemical composition appears to account for a large number of cases of dolomitization in modern and ancient sequences (Fig. 4.9 C, D). The slow crystallization resulting from the mixing of two different interstitial solutions leads to clear dolomite in sparry well-formed crystals.

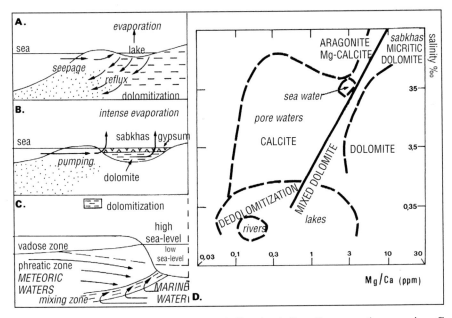

Fig. 4.9 A–D. Models of dolomitization. **A** Infiltration/reflux; **B** evaporative pumping; **C** mixing of continental and marine pore waters; **D** fields of precipitation of principal carbonates in relation to salinity and Mg/Ca (after Folk and Land 1975)

d) The model of late diagenetic dolomitization and ankeritization through the transfer of magnesium liberated from Mg-calcite, and from clay minerals during transformation of smectite to illite, appears to be applicable to limestones with varying marl content that have been deeply buried.

e) Let us recall that the process of dedolomitization also occurs, especially where dolomitic limestones are pervaded by meteoric waters rich in SO_4. Such waters may result from the dissolution of sulfate-bearing evaporites or of oxidized pyritic marls.

4.5 Evolution of Siliceous Deposits

Deep-water siliceous deposits are frequently encountered in Mesozoic and Cenozoic oceanic sequences (Hsü and Jenkyns 1974; Warme et al. 1981). Planktonic tests, consisting of amorphous opal A, are transformed initially into opal CT (unordered structure of cristobalite/tridymite-type), then into chalcedonic microquartz, and eventually into quartz (Calvert, in Hsü and Jenkyns 1974; Dapples, in Larsen and Chilingar 1979; Riech and von Rad 1979; Pisciotto, in Warme et al. 1981). This transformation, by way of dissolution and reprecipitation, leads to the destruction of the original organic structures, and to the successive formation of microcrystalline aggregates (lepi-

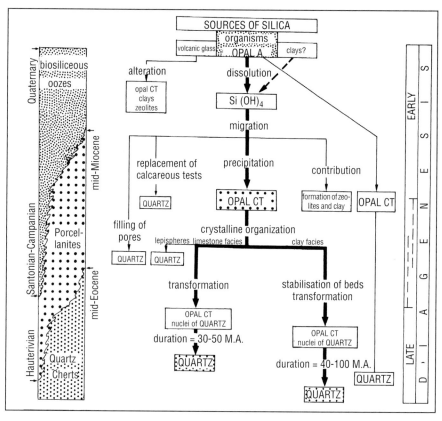

Fig. 4.10. Modes of siliceous burial diagenesis, according to data from the North Atlantic Ocean (after Riech and von Rad 1979)

spheres) or granular cement (lussatite), to porcellaneous cement (porcellanite), and then to massive quartzose cherts (Fig. 4.10). There are notable differences in this diagenetic sequence between more calcareous sequences, in which silica changes rapidly to quartz, and more clayey successions, in which a slower transformation of lepispheres to opal CT takes place. Thus, Mesozoic and Cenozoic oceanic sediments contain well-crystallized quartzose cherts in calcareous turbidites, and lepispheres of opal CT together with well-preserved radiolarians in the fine-grained clay-rich pelagites. These differences appear to be controlled by the lower permeability of argillaceous layers and the more widespread presence of impurities in them which lead to a less well-ordered crystal growth from the siliceous nuclei (cf. discussion in Leeder 1982).

Siliceous shallow-water deposits are frequently abundant in ancient geological series such as the lower Carboniferous limestones and the Cretaceous chalks in Europe, or the lacustrine Tertiary limestones of the Paris basin where they occur as layers of nodular chert or flint. It is agreed that the diagenetic formation of these cherts can take place through one of two processes.

a) Stationary interstitial waters rich in silica resulting from the solution of tests within porous beds permit the precipitation of this compound around growth centers. The restriction of these waters to certain levels within the rocks may result from the presence of phreatic horizons, or from the inhibition of downward infiltration above comparatively impermeable layers.

b) Mixing of pore waters of meteoric and marine origin in proportions leading to an undersaturation against calcite and an oversaturation against opal, may result in precipitation of the latter. This model conveniently explains the presence of layers of siliceous nodules in numerous neritic or continental shallow-buried limestones. In principle, it is similar to the model invoked for dolomitization in infratidal environments (Sect. 4.4.3; Fig. 4.9 C). Note that the outer layer or cortex of the chert or flint nodules is usually not the result of secondary meteoric alteration, but represents the final stage of the diagenetic chertification.

4.6 Formation of Fossil Fuels

4.6.1 Coal

Coals result from the progressive physico-chemical evolution of plant debris. The original sedimentary environments are characterized by prolific plant growth, reduced supply of terrigenous matter, a fairly rapid burial preventing oxidation, and the biological degeneration of the dead tissue. Such environments are developed in swamps, lagoons, and moors in alluvial valleys, along the coastal plains, deltas, and back-reef zones in humid tropical regions. Modern examples are the mangrove swamps along the coast of Florida and the Gulf of Mexico (cf. Galloway and Hobday 1983) or large deltas such as the Mahakam on the Indonesian island of Kalimantan (Allen et al. 1979). The majority of coals, the so-called *humic coals,* are finely bedded and result from the evolution of peat from macroscopic plant debris during which the lignin is transformed by microbes into humic acids and various types of residues. The unstratified, less common *sapropelic coals* result from the subaqueous accumulation or organic oozes rich in debris derived from algae or pollen. The various types of peat, coal, and graphite are ranked according to the enrichment of carbon and the depletion of the volatile matter in the course of diagenesis (Table 4.2). The various types of plant tissue or *macerals,* the principal compounds of coal, are subdivided into the most common vitrinite of humic origin, liptinite (=exinite) rich in lipids, and into inertinite consisting of hard fragile debris (Table 4.3). In addition to the vegetal remains, coals also contain various sedimentary compounds, the abundance of which largely determines their economic value. Amongst these materials are quartz, clay,

Table 4.2. Principal features of the diagenetic evolution of coal and hydrocarbon deposits (after Tissot and Welte 1978; Perrodon 1985)

Stage	Vitrinite reflectance	Coals Rank		%C (vitrinite)	% volatile matter	Petrographic characters	Hydrocarbons
Diagenesis	0.3	Peat		50		Free cellulose, plant details recognizable	Methane and early gases (immature zone)
		Lignite:	soft dull brilliant	60		Free cellulose absent	
Catagenesis	0.5	Coal:	high-volatile	75	45 40	Clean, clearer liptinite ("coalification step")	Oil ("oil window")
	1		medium-volatile		30		
	1.5				20		
	2		low-volatile		10		Moist gas
Metagenesis	2.5	Anthracite		90	5	liptinite indisting vishable from vitrinite	Late methane (dry gas)
Metamorphism	4	Graphite		100		Anisotropic reflectance	?
	11						

Table 4.3. Main maceral groups of coal deposits (O oxygen; H hydrogen; C carbon)

Maceral group	Macerals	Origin	Lithotypes
Vitrinite (rather abundant O)	Collinite Telinite	Small-sized plant debris wood, bark	Vitrain
Liptinite (abundant H)	Sporinite Cutinite Resinite Alginite	Spores Cuticules Resin Algae	Clarain
Inertinite (abundant C)	Micrinite Sclerotinite Semifusinite Fusinite	Fine organic particles Fungi Lignitic tissue Homogeneous lignitic tissue	Durain Fusain

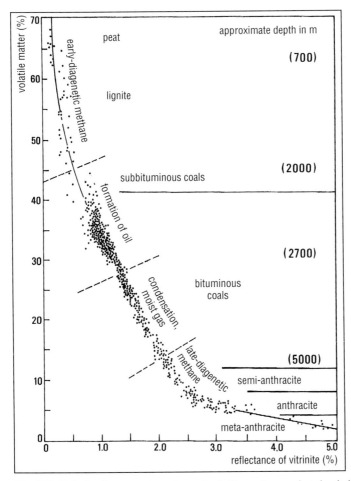

Fig. 4.11. Relation between percentage of volatile matter, coal rank, vitrinite reflectance, and types of hydrocarbons in the Carboniferous of Germany (after Stach, and Teichmüller and Teichmüller, both in Blatt 1982)

detrital heavy minerals, concretions, and individual minerals resulting from early diagenesis, such as carbonates, sulfides, and subspherical nodules around calcitized or dolomitized debris, the so-called coal balls, etc.

The diagenetic evolution of coals (Fig. 4.11) is mainly controlled by the increase of temperature during burial, leading to

1. lignites down to 1000 m or more;
2. coals with increasing contents of bituminous matter down to 5000 m at temperatures of 100°–200° C (deep diagenesis = catagenesis);
3. anthracites below about 5000 m (anchizone = metagenesis) (cf. Teichmüller and Teichmüller, in Fairbridge and Bourgeois 1978; and in Larsen and Chilingar 1979; Tissot and Welte 1978).

The evolutionary state of a coal is measured by the reflectance of the macerals, i.e., the ratio between the reflectance of a standard incident monochromatic green light from a certain maceral and from a known reference surface. The reflectance which is directly correlated with the content of volatile matter, water, and certain chemical elements (C, O, H, N), is generally determined on vitrinite, as this maceral is also present in numerous rocks other than coal. It thus presents a precise parameter of the diagenetic stage attained by a rock (Sect. 4.2.3; Figs. 4.3 and 4.11). The increasing maturity of coals, favored by greater depths and pressures, leads to an enrichment of diagenetic methane and to a progressive elimination of the denser hydrocarbons such as hexane and pentane, in favor of the lighter hydrocarbons.

4.6.2 Hydrocarbons

Petroleum and natural gas result from the transformation of small amounts of organic matter in fine-grained sedimentary source rocks. In shales their mean concentration is about 2.1%, in carbonates 0.29%, and in sandstones 0.05% (cf. Tissot and Welte 1978; Perrodon 1985). The organic matter is derived mainly from plankton and soils, and the main rock type in which it is contained is shaley and of dark color, i.e., the black shales in a wider sense. Formation and preservation of organic matter are controlled by elevated productivity, a deficiency in oxygen in the bottom waters and interstitial solutions, and rapid burial (Galloway and Hobday 1983). During diagenesis, the various organic components (biopolymers) derived from plants or organisms are transformed into geopolymers called *kerogen,* from which oil and gas are then formed (Fig. 4.12). The evolution starts under cover of only a few tens of meters with an early diagenetic degradation by methanogenic bacteria, leading to the formation of amino acids and sugars. These products in part are consumed by the microorganisms, and the remainder is recombined through polymerization and condensation to fulvic acids and brownish humic complexes to form kerogen. During burial the kerogen develops further through thermal degradation. Oil originates at rather moderate temperatures, whereas

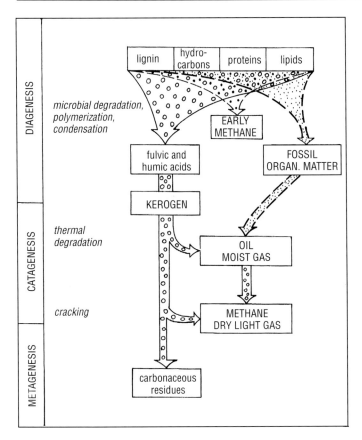

Fig. 4.12. Main stages in the development of hydrocarbons (after Tissot and Welte 1978)

methane-bearing gases, becoming lighter and drier with depth, are formed deeper down, through the progressive modification of the chemical structures of the hydrocarbons, a process referred to as cracking (Figs. 4.11 and 4.12; Table 4.2). Average conditions for the formation of petroleum are encountered at a depth of about 3000 m and temperatures of 50°–150° C in the "oil window", beyond which natural gas will form, mostly between 120°–200° C. The exact characterization of the diagenetic formation of hydrocarbons is done with the aid of the types of kerogen, as defined by their content of hydrocarbon and oxygen, compared to their carbon content (Fig. 4.13). Type I kerogens are rich in lipids predominantly of algal origin (alginite, Table 4.3), and are the source of abundant oil deposits. Type II kerogens are particularly good sources of oil, rich in liptinite and alginite mainly from organic matter of marine origin. Type III kerogens are derived from material of higher land plants (vitrinite), have a low oil potential (more gas), but are resistant to alteration.

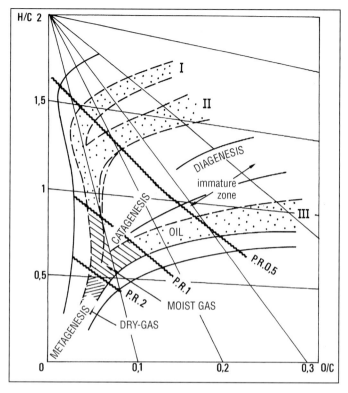

Fig. 4.13. General evolutionary scheme of kerogen, after diagram of van Krevelen (after Tissot and Welte 1978; Perrodon 1985); *I, II, III* types of kerogen; *PR* vitrinite reflectance (see text for details)

In order to become concentrated in economic amounts, the hydrocarbons formed during diagenetic evolution have to be expelled from their fine-grained source rocks to the coarser-grained reservoir rocks of higher porosity (*primary migration*), and then transported into geological traps from which they cannot escape (*secondary migration*). Primary migration takes place over only fairly short distances into coarser-grained well-sorted rocks which may have formed in a variety of environments: river channels, desert dunes, platform carbonates and reef zones, sandy littoral zones (delta fronts, near-shore detrital wedges), and detrital submarine fans. Primary migration appears to take place fairly soon after the phase of active oil formation (cf. Perrodon 1985). It seems to be favored by the diagenetic transformation of smectites to illites which tends to be accompanied by a liberation of water into the pore volume, a relative volume reduction due to compaction, and a slight increase in porosity (Bruce 1984).

The transportation of hydrocarbons can take place in true solution, in colloidal solution, or in the form of individual gas or oil bubbles. It is facilitated

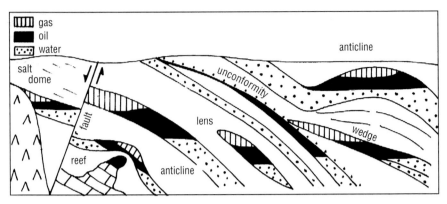

Fig. 4.14. Types of hydrocarbon traps (after Leeder 1982)

by elevated porosities. Detailed knowledge of the processes of silica and carbonate cementation is of fundamental importance in petroleum geology (cf. Sects. 4.4.2 and 4.4.4). The most favorable reservoirs are characterized by weak cementation after migration of the hydrocarbons has taken place, as in calcareous sandstones of the Arabian Peninsula, or by secondary porosity resulting from dissolution at depth such as in the Devonian reefs of Canada. Other favorable features are late fracturing, e.g., in the reservoir rocks of Iran and Iraq, as well as in certain Cretaceous limestones of the North Sea.

The secondary migration of hydrocarbons into various types of traps may require a long time and, due to the differences in density, results in a layering which is the inverse of the depths of formation, i.e., gas at the top and water at the bottom. Such structures are mostly anticlinal, but are also formed by faults, unconformities, sedimentary wedges, reefs, etc. (Fig. 4.14).

There are also a number of large hydrocarbon deposits which are difficult or uneconomic to exploit. Such deposits include the tar sands of the Athabasca Formation of western Canada. These formed at temperatures of about 90° C along the contact between oxygenated meteoric waters and oil, and by solution and biodegradation of the lighter compounds. This is also the case for the bituminous shales of the Eocene Green River Formation of the western USA and certain Kimmeridgian shales of northwestern Europe which resulted from the prolific growth of phytoplankton, the so-called blooms, and are rich in oils of algal origin.

5 Continental Sedimentation

5.1 Glacial Environments

5.1.1 Introduction

The present *glacial cover* extends over about 10% of the surface of the earth, containing some 75% of its total fresh water resources. During the Quaternary maximum glacial periods, however, about 30% of the land surface was covered by ice caps. At least five major glacial periods occurred in the geological past: the lower Proterozoic in North America; the upper Precambrian on virtually all continents with the exception of Antarctica; the Ordovician of Africa and South America; the upper Paleozoic on Gondwanaland of the southern hemisphere; and the upper Cenozoic, starting with the Oligocene and with maximum development in the Pleistocene, in high latitudes and mountain ranges of the temperate regions (cf. Flint 1971; Goldthwait 1975; Frakes 1979). The glaciations can be ascribed mainly to cyclic fluctuations in the quantity of solar energy received at high latitudes as a result of changes in the inclination of the earth's axis (Hays et al. 1976; see Kennett 1982). The chance of sediments derived from glacial erosion and transport to become preserved is rather small, like that of the sedimentary particles imprisoned in the ice. The effects of reworking by moving ice and the associated gravitational phenomena are little conducive to the preservation of such glacial deposits. Detailed glacial "archives" are available on a variety of scales, for example, on subsiding platforms close to old continents (West Africa, Boeuf et al. 1971; Deynoux 1978) or in the thick glacial carapaces of higher latitudes (Antarctica, Greenland; Eyles and Miall, in Walker 1984).

The large glacial masses form mainly in valleys, at the foot of mountain ranges, as inland ice, and in ice shelves (Fig. 5.1). Glaciers represent an open environment in which active supply of ice from upland areas and its downhill merging result in a fluctuating equilibrium. In *polar glaciers,* the temperature is always much below the freezing point of water. There is thus little or no subglacial drainage, erosion taking place solely by planing away the surface rocks beneath the glaciers. As there is also no transporting liquid, transport takes place in rigid form. In the thick glaciers of *temperate regions,* formed independently of latitude, the temperature at depth approaches the melting point of

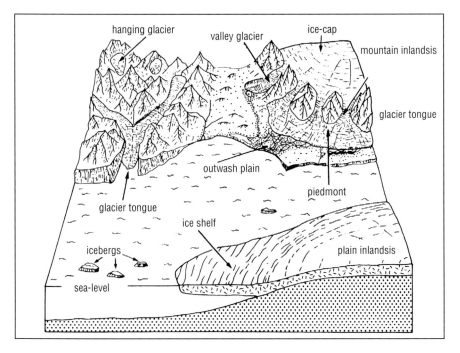

Fig. 5.1. Main glacial environments (after Allen 1977)

ice due to the intense pressure. The resulting plastic basal ice slides down-valley on its own substrate, leading to intense deformation of the underlying material. The bottom waters, derived from seasonal variations in temperature as well as from the increase of geothermal gradient and pressure with depth, are abundant below the glacier. Charged with debris, they combine forces with the ice itself in causing very strong and complex erosion in the form of abrasion, crushing, plucking, and fracturing. Particles plucked from the substrate may accumulate in seasonal layers which become deformed over the years. They can also rise to the surface of the ice or gather in overlapping layers in the downflow direction.

5.1.2 Main Types of Glacial Deposits

The deposits formed by ice result essentially from mechanical action or physical weathering and are made up of mixtures of rock fragments and minerals, ranging from giant boulders to extremely fine rock dust or so-called rock flour, all abraded from the substrate. The various constituents of glacial or fluvio-glacial deposits faithfully reflect nature and origin of the substrates, sometimes even for periods as old as the upper Precambrian or the Paleozoic

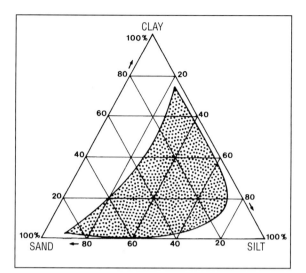

Fig. 5.2. Grain-size distribution of clayey-sandy matrix of North American moraines (range of more than 1000 samples of different origin) (after Dreimanis, in Fairbridge and Bourgeois 1978)

(Chamley et al. 1980). The particles are mostly angular, heterogeneous, and poorly graded (Fig. 5.2), frequently striated, grooved, and faceted by the force of friction and pressure. The substrate is deeply worn by the actions of the ice into U-shaped valleys, and present a rounded, frequently smoothed surface commonly with striae and abrasion structures, and residual hummocks or "roches moutonnées".

Truly glacial sediments are represented by *moraines or tills* (Table 5.1), in which the water content at the time of deposition is too low to permit sorting or segregation. Below the glaciers of temperate regions, basal or ground moraines or block tills are developed in the form of sedimentary layers of variable thickness and great extent. They are highly diverse in particle size (diamictites), non-stratified, and consist of boulders engulfed by a clayey-sandy matrix derived from abrasion. Boulders, pebbles, and sand grains are preferentially aligned with their larger side parallel to the transport direction. A smaller portion is arranged at right angles to this direction under the influence of shearing and by violent subglacial water currents. The basal tills sometimes occur in elongated masses called drumlins, especially on the large cratonic regions once covered by ice sheets. Lateral moraines are developed along the sides of a glacier and medial ones result from the confluence of individual glaciers. Terminal moraines are pushed in the down-flow direction, and summit or supraglacial moraines result from block falls onto the top of the ice. These latter types of moraines are usually thinner, less compacted, coarser, and sometimes better graded than the basal moraines. On the continents, lines of terminal moraine ridges mark the stages of maximum glaciation. Sometimes the continued melting of a glacier leads to excessive accumulation of material derived from the ice which then is susceptible to downward sliding giving rise to a flow till. The heterogeneous appearance of

Table 5.1. Main glacial environments and sediments

Depositional environment	Sedimentary facies	Agents	Topography
Glacial (s. str.)	Massive mixtites, poorly sorted, with sandy-argillaceous matrix, unstratified, compacted = basal moraine or till	Live ice, moving slowly downslope	Smoothed surfaces, more or less rounded topography, elongated drumlins
	Mixtite of variable thickness, frequently poorly sorted = lateral, frontal, medial, supraglacial moraine or till	Live moving ice, melting, gravity	Lobes, rings, boulder belts, superimposed massses
Fluvio-glacial	Sand, pebbles, boulders, medium to rather well sorted, thickly bedded or channels. Better sorting in downflow direction	Water in contact with ice: on, within, or below the glacier; melting and intense currents. Torrents directly issuing from glacier	Eskers (= sinuous sand stringers) Kames = mounds Kames terrace (along border of glaciers) Sandurs (= outwash sheets) Valleys with torrential flows, frequently incised
Glacio-Lacustrine	Rhythmites of silt or clay (varves)	Seasonal discharge	Borders of glacial outwash plains, Lacustrine basins
Glacio-marine	Irregular laminites of fine-grained sediment, with isolated pebbles, frequent resediments (turbidites, debris flows)	Marine transport in agitated environment, as suspensions or by floating ice	Continental shelf and slope
	Massive mixtites, unlaminated, rich in clay and silt, with isolated pebbles (= aquatill)	Marine transport in quiet environment, as suspensions or by floating ice	Open-marine sea floor

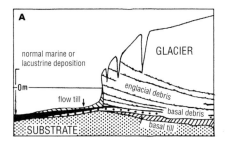 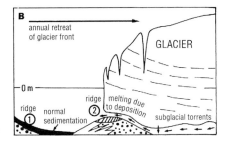

Fig. 5.3 A, B. Sedimentation at the contact between a glacier and a body of water (sea, lake) (after A. Dreimanis, and I. Bannergee and B. C. McDonald, in Leeder 1982). **A** Formation of a subaqueous moraine (or till) due to melting of ice and collapse of the equilibrium in the sedimentary material accumulated; **B** formation of successive seasonal mounds (*1, 2*), accumulating as the result of the slow-down of the subglacial waters issuing from below the retreating glacier

the resulting deposits is reminiscent of a debris flow (see Chap. 3). This convergence becomes truly amazing when the march of the moraine ends in a subaqueous environment when the glacial tongue is located close to a lacustrine or marine beach (Fig. 5.3 A).

The fluvio-glacial sediments deposited within the glacial masses or in their immediate vicinity, consist essentially of sand-sized or coarser particles, as silt and clay are carried away downstream in suspension. They are moderately graded, and frequently show cross-bedding, climbing ripples, and scour marks. Eskers are sedimentary bodies of positive relief extending over several kilometers in length at widths of 50–500 m and heights of 5–50 m, elongated in the direction of transport. They result from the deposition of particles from sediment-rich extremely violent subglacial torrents, especially in front of the glaciers during phases of retreat. They can exhibit a slope opposite to that of the enclosing valley because of the hydrostatic pressure under the weight of the ice above, e.g., a glacial ridge.

The arrival of such a torrent in a subaquatic environment can lead to a loss of competence and to bulk deposition of particles in the form of successive ridges during the seasonal retreat of the glaciers (Fig. 5.3 B). Subglacial accumulation in the form of isolated hills, or kames, corresponds to zones of eddies, glacial funnels, or pronounced confinement of water flow along the sides of glaciers in kames terraces. They are widely developed in zones of abrasion in the lower reaches and consist of moderately to well-sorted sediment frequently associated with ice-created impact marks. Alluvial fans situated below the glaciers, are supplied with sandy-gravelly material by laterally diverging torrents (cf. Sect. 5.4). They result in sedimentary outwash cones or sandurs, in which the fineness of the material, as well as sorting and bedding become more pronounced in the downstream direction. The same type of sedimentary gradation occurs in the fluvio-glacial alluvials deposited by more rectilinear water courses connecting the glacial environment with the true fluvial systems farther downstream.

136 Continental Sedimentation

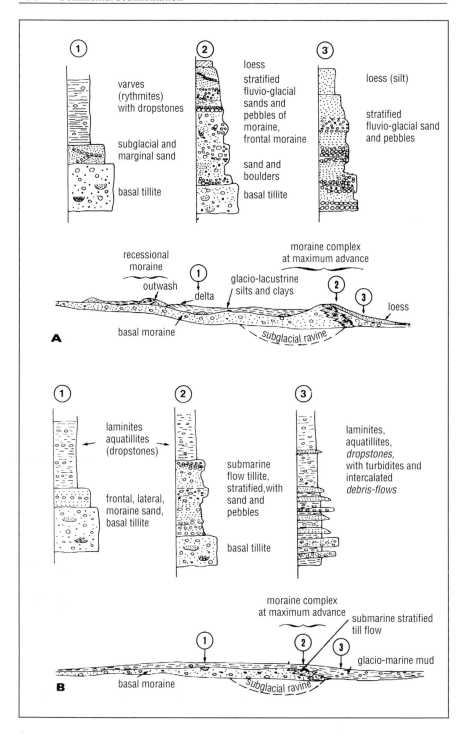

Glacio-lacustrine sediments are generally of fine grain size, save for the rare marginal deposits of a more varied nature, for example, sand bodies shaped by currents, or pebbles and blocks derived from older moraines eroded by waves, etc. Glacio-lacustrine deltas present a greater variety of structures than classical deltas (cf. Chap. 6). This is due to the heterogeneity of the feeding glacial environments, and to fluctuations in energy level. The usual deltaic sequence of sandy-silty stratified sediments (bottomset, foreset, topset; cf. Sect. 6.1.1) can be disturbed by deposition from floating ice, from the rupture of subglacial water pockets or lakes, and from outwash tills. The bottoms of glacial lakes are characterized by alternating fine laminae (rhythmites) caused by seasonal variations in the supply of detritus: sandy light-colored silts during summer, and dark organic clays in winter (Mörner, in Fairbridge and Bourgeois 1978). These so-called varves, resulting from settling in a nonflocculating environment, are locally disturbed by dropstones derived from melting floating ice or by turbidity currents. Below glacial lakes, torrential fluvio-glacial deposits may be formed, as well as loess, unstratified clay-depleted silty deposits, the result of deflation over the outwash plains at the foot of the glaciers (Fig. 5.4 A).

Glacio-marine deposits are distinguished from lacustrine ones by the absence of varves. They consist of silty clays with coarse lamination or bedding as unordered laminites, or in homogeneous bodies as aquatills. They are mostly light-colored and contain dropstones of various sizes. In the direction of the coast they frequently grade into subaqueous outwash tillites, whereas seaward, away from the influence of the pack ice, they are interrupted by debris flows and turbidity currents (Fig. 5.4 B). Farther out, the glacio-marine muds gradually pass into normal oceanic sediments of the continental margin, dominated by hemipelagites intercalated with various types of resediments.

5.1.3 Glacial Sequences and Environments of the Past

Ancient glacial deposits tend to be best preserved when they have been buried rapidly, and in subaqueous environments they can attain considerable areal extent (Edwards, in Reading 1986). The glacial sequences are usually made up of a number of distinct entities: firstly, the basal erosion surface, secondly the basal moraine contemporaneous with the advance of the glacier and subsequently consolidated to become a tillite or morainite, and eventually deposits resulting from surface moraines and periglacial deposits formed during the retreat of the glacier (Fig. 5.4). The associations of typical sedi-

Fig. 5.4 A, B. Schematic sections and details of sedimentary sequences observed after the retreat of a glacier in terrestrial (**A**) and marine (**B**) environments (after M. B. Edwards, in Reading 1986); *1* internal facies association (upslope or coastal environment); *2* marginal facies; *3* external facies (downslope, open marine)

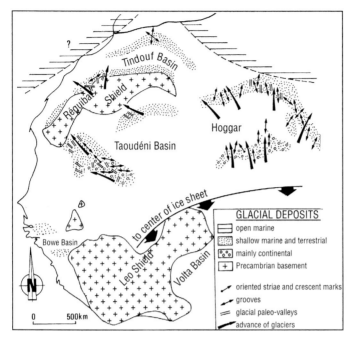

Fig. 5.5. Paleogeographic reconstruction of the upper Ordovician glaciation in West Africa (after Deynoux 1985)

ments from glacial environments identified in ancient sequences can be useful indicators for paleogeographic reconstructions. In *terrestrial environments,* internal structures of basal moraines or tills can be recognized, as well as marginal structures in lateral or supraglacial moraines, and external structures in frontal moraines or aeolian deposits. Retreat of the glaciers leads to formation of lacustrine rhythmites and outwash plains behind the abandoned terminal moraines, such as observed in the Alps (Wright 1937). In the *marine environment,* disordered laminites and unstratified aquatills with dropstones predominate. They are developed over thick basal tills in the internal association, over coarse subaqueous outwash material and thin basal tillites in the marginal associations, and intercalated with marine resedimented material in the external ones like the Weddell Sea of Antarctica (Anderson et al. 1977). Paleogeographic reconstructions can be prepared with the aid of the spatial arrangement of glaciogenic structures and sediment bodies in sequences ranging from the Precambrian to the lower Paleozoic (Fig. 5.5) as exemplified by Boeuf et al. (1971) and Deynoux (1985). Note that lithological variability of glacial deposits is reduced under certain geographic and climatic conditions. For instance, the marine facies developed in high latitudes during the long cold periods of the Pleistocene are characterized by large thicknesses of monotonous fine-grained muds with dropstones (Ross Sea, Gulf of Alaska, North Sea; Caston 1977).

5.2 Deserts

5.2.1 Introduction

Aeolian processes predominate in areas in which the annual precipitation is below 250 mm, and at the same time deficient in comparison to evaporation at high temperatures. This explains the importance of tropical deserts which occupy some 20% of the surface of the earth between 10°–30° latitude in the trade-wind belts (Fig. 5.6). Desert zones are also developed outside the tropics wherever precipitation is held back by mountain ranges or in the vicinity of large basins with marine evaporation. Only about 20% of the desert surfaces are covered by aeolian, mostly sandy deposits. The remainder are made up by denuded zones over which physical weathering and erosion supply material for deflation, as well as by temporary water courses and lakes (cf. Glennie 1970; Cooke and Warren 1973; Reineck and Singh 1980; Brookfield, in Walker 1984). At times during the Pleistocene the areal extent of deserts was larger than at present, indicating an increased importance of the trade winds, of upwellings of deep waters along the continental margins, and of the transfer of sand from land to sea (Sarnthein et al. 1980).

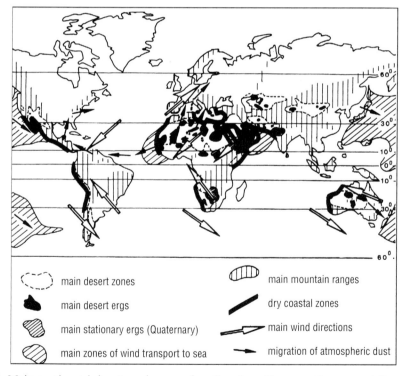

Fig. 5.6. Main continental deserts and zones of aeolian deposition over the oceans (after Cooke and Warren 1973; Glennie and Folger, in Fairbridge and Bourgeois 1978)

The vast accumulations of dunes which contain some 90% of the present mobile aeolian sand reservoir, are controlled by the position of the main wind zones and atmospheric pressure systems. The large "flows" of sand like those advancing across the Sahara (Wilson 1971), are distributed in drainage basins comparable to fluviatile basins, where the peaks correspond to high pressure areas, and the valleys to lower pressure areas. Direct relations with topography are always rather limited, as the wind-driven sand flows easily cut across the actual relief. The large aeolian fluxes of sand frequently are directed to the edge of the continents and thus play an important role in marine detrital sedimentation. This is in particular the case for the northern Pacific, the sea to the east of Australia, and the northwestern part of the Indian Ocean (Fig. 5.6). The situation also applies to the Atlantic Ocean off the Sahara where the aeolian contribution to marine sedimentation can amount to about 260×10^6 t per year (Schütz and Jänicke, in Sarnthein et al. 1980).

5.2.2 Recent Desert Sediments

Erosion by wind leads to the accumulation of sand in the deserts, whereas silt and clay are carried into the surrounding continental areas and the marine basins to form loess and other deposits of aeolian dust. Farther away from the sources, the grain size tends to decrease and the petrographic spectrum is reduced in favor of the more resistant components. More fragile minerals, such as the fibrous clay palygorskite, are always prone to covering long distances, especially when made up of small aggregates and transported by high-altitude winds like volcanic dust (Coudé-Gaussen 1982). Sand grains resulting from wind action are characterized by rounded outlines and frosted surfaces, their grain size varying between 0.1–1 mm, depending on the locality. They exhibit very good sorting with positively skewed size distributions, and an abundance of quartz grains of great textural maturity, conducive to the formation of quartz arenites. The presence of pebbles with faceted surfaces separated by finely rounded edges, notably in the form of "dreikanters", i.e., with three facets, and of boulders with sculptured and grooved surfaces is particularly helpful in recognizing aeolian deposits. With the exception of bones and vertebrate tracks, fossils are generally rare. Ripples formed by wind action are asymmetric, with rectilinear, sometimes bifurcating crests, and show higher ripple indices (wavelength/height). The direction of their crests can diverge significantly from that of the wind forming them and is usually more variable than in the case of subaqueous ripples. The cross-beds are highly inclined against each other and frequently rather continuous. The beds do not possess silty-clayey intercalations and form units up to several meters thick. Sand avalanches down the lee side of the ripples are rather irregular, leading to poorly discernible, irregular cross-laminations. These characteristic features facilitate the recognition of aeolian ripples (cf. Glennie 1970). *Aeolian dunes* resulting from the *accumulation* of sand make up the following main types (Fig. 5.7):

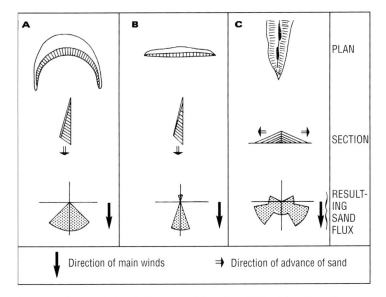

Fig. 5.7 A–C. Main types of dunes (after Galloway and Hobday 1983): **A** barkhan; **B** transverse dune; **C** longitudinal dune

1. barchans of lunate shape with the convex side facing the wind;
2. transverse dunes oriented at right angles to the dominant wind direction;
3. longitudinal dunes or seifs, oriented parallel to the wind and resulting from bilateral helicoidal movement of air leading to a symmetrical internal structure.

In addition to these main types which occur on different scales, there are also the small nebkas extending from behind small obstacles such as knolls or individual boulders, and the parabolic dunes, the hollowed-out center of which appears to result from local acceleration of the wind velocity. The complex combination of these various structures constitutes the ergs or sand seas in the large deserts.

The zones of accumulation in deserts form concurrently with *zones of ablation,* of which there are also a number of types: denuded plateaus or hamadas; surfaces strewn with fragments too large to be moved by wind, the regs; and isolated massifs of resistent rocks in the form of inselbergs. Deserts, however, also contain a number of sedimentary facies resulting from the action of water. Ephemeral streams carry along pebbles, sand, and clay which they leave behind in long surficial ribbons or in wadis as soon as the water seeps into the ground and dries off, after the cessation of the rain. Sedimentary structures observed in these fluvial deposits are indicative of the upper and lower flow regimes (cf. Chap. 3). The surface of these deposits frequently is covered by mud cracks, the clay flakes of which can be covered by sand or carried away by wind. The sedimentary sequences observed at the interface between aeolian and fluvial influences are rather complex (Fig. 5.8). The pediments

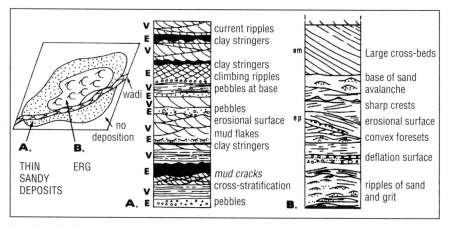

Fig. 5.8 A, B. Sedimentary sequences resulting from truly aeolian processes in an erg (**B**), and by the combination of aeolian and fluvial processes along its margin (**A**), (after Fryberger et al. 1979); *V* wind action; *E* water action; *ep* precursor of erg; *em* mature erg

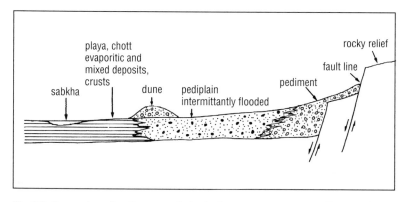

Fig. 5.9. Succession of sedimentary facies in desert country surrounding a rocky massif (after Friedman and Sanders 1978)

developed below inselbergs are reworked by floods resulting from downpours during thunderstorms. They form dense discharge flows with an clayey matrix in the form of conglomeratic mudflows similar to debris flows. These pediments grade downslope into large subhorizontal pediplains or playas (Figs. 5.9). In depressions on these surfaces, ephemeral lakes may establish themselves, in which evaporation of rain-water leads to precipitation of halite and gypsum, encrustating the fluvial and aeolian deposits. The crystallization of the evaporitic minerals results also from the tapping of phreatic water horizons in the sabkhas, saline soils or chotts, and in the surrounding dunes, as well as from capillary phenomena controlled by diurnal or seasonal fluctuations in temperature. Other features resulting from the vertical migration of water within sands are carbonate crusts and so-called sand roses with brown-red coatings of mostly ferric oxides so characteristic of ancient dune sands.

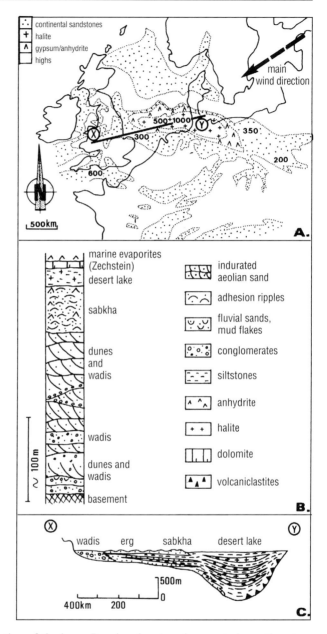

Fig. 5.10 A–C. Reconstruction of the lower Permian desert environments in northwestern Europe (after Glennie 1972; Ziegler 1975). **A** Paleogeographic situation (*figures* show thickness of strata); **B** schematic sedimentary sequence; **C** reconstructed section through the basin during the Rotliegend deposition

5.2.3 Ancient Desert Sediments

Ancient aeolian sediments are mainly distinguished by the following features (Glennie, in Fairbridge and Bourgeois 1978): rounded and frosted sand grains, good sorting, and depletion in argillaceous matrix to below 5%; crossbeds oriented at different, sometimes high angles of up to 34°; inclined planar bedding with few recognizable ripples because of frequent truncations by wind action; quartz is extremely abundant and often shows ferruginous coating, mica is generally absent. Ancient sediments deposited by water in deserts are distinguished from those of other environments by a number of characteristics encountered individually or in combination with each other: abrupt upward decrease of grain size resulting from the rapid drop in transport energy of the waters which become absorbed by the sands after a rainfall; presence of mudstones with isolated pebbles as well as stringers of pebbles and mud chips; mud cracks and sand-filled clastic dykes occupying the fissures; and adhesion ripples forming just prior to emersion. Other characteristics are the presence of calcitic cement or encrustations, ferruginous coatings, and locally gypsum or anhydrite as cements or encrustations.

Regional reconstructions of desert environments in ancient sequences are rather rare (cf. Glennie 1970; Collinson, in Reading 1986). One of the better-established examples are the red sandstones of the lower Permian or Rotliegend of northwestern Europe and the southern part of the North Sea. Their deposition started on the Variscian basement, and was terminated by the marine transgression leading to the evaporitic Zechstein formations. Borehole data revealed a 500-m thick fossil erg complex, bordered by fluvial facies with wadis, and lacustrine facies with sabkhas (Fig. 5.10). Lateral migration of the aeolian bedforms under the influence of the predominantly north-easterly winds could be reconstructed from sedimentary structures and facies sequences. The moderately cemented fossil dune sands represent an important reservoir rock for natural gas in the Netherlands and the North Sea. The porosity is reduced, however, at the level of intercalated heterogeneous fluvial deposits.

5.3 Lakes

5.3.1 Introduction

5.3.1.1 Location

Recent lakes occupy only a relatively small portion of the earth's surface. Nevertheless, they represent a highly complex array of locations, surface extent, depth, as well as types of waters and sediments. There is every intermediate stage between the large, permanent and deep freshwater lakes like those of the East-African Rift, or North America, and Lake Baykal of Siberia, and the ephemeral, saline, and shallow lakes of Iran or the western United States. Deep lakes also may be saline, like the Dead Sea or the Caspian Sea, just as shallow lakes may contain freshwater like Lake Chad. The fossilization of lacustrine sediments is supported by active subsidence resulting in strong sedimentation and rapid burial, as in lakes forming in the incipient rift zones of East Africa, the Dead Sea, or Lake Baykal. The life expectancy of such lakes in zones of expansion is frequently shorter than that of lakes forming in depressions on stable cratonic crust, like Lake Chad or Lake Eyrie of Australia. Compared to their recent equivalents, fossil lakes may exhibit considerable areal extent (Lerman 1978; Matter and Tucker 1978; Picard and High 1981). This is the case for the Triassic Lake Popo Agie of Wyoming (130 000 km^2) or the Pleistocene Lake Dieri of Australia (110 000 km^2). Of the recent lakes, only the Caspian Sea covers a large area (375 000 km^2), followed by Lake Victoria with 70 000 km^2. The thickness of ancient lacustrine sediments can be considerable, reaching about 4000 m in the Devonian of Scotland. In comparison to this, sediments in recent lakes are rather thin, about 200 m in Lake Geneva and 20 m in Lake Eyrie.

The characteristic properties of lacustrine waters and sediments are mainly controlled by:

1. climate, which influences the mean water level, the chemical composition of the waters, their temperature, and organic productivity;

2. thickness of the water layer which affects currents and layering of the water mass; and

3. peculiarities of the drainage basin which determine nature and quantity of the particulate and dissolved freight of its rivers and streams.

5.3.1.2 Water Dynamics

Movements of lacustrine waters are controlled by winds and by currents of fluvial origin, as well as by seasonal fluctuations in water density. The surface area of lakes usually is too small to become influenced by tides. Winds tend to propel the surface waters, and on larger lakes the advance is deflected by the Coriolis force. The water level tends to rise in the direction of beaches exposed to the winds, whereas ahead of the zone of direct wind action waves are low and the water level is lowered.

Seiches are strong waves of great amplitude and up to several meters high which result from the back-flow of waters pushed by wind-generated waves against the lacustrine beaches. Isolated waves of this type are known from Lake Geneva as a periodic phenomenon.

The majority of lakes exhibit pronounced *thermal stratification,* in which the warmer surface layers are separated from the colder deeper waters by a discontinuity referred to as the thermocline. This stratification varies little in deep tropical lakes, but shows pronounced seasonal fluctuations in lakes of temperate regions. The oxygenated surface waters, the epilimnion, are warmed up during spring until they represent a relatively lighter mass during

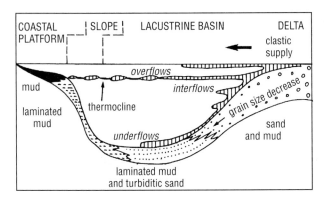

Fig. 5.11. Model of detrital sedimentation in a temperate lake showing thermal stratification and subjected to fluvial supply of material (after Sturm and Matter 1978)

summer. This inhibits the vertical exchange of waters and the descent of suspended particles to the bottom. The thermocline is firmly established and leads to a retention of some of the descending particles at this level, and to an oxygen deficit in the underlying body of water. During autumn, the surface waters gradually cool off and become specifically heavier, leading to mixing with the deeper waters, the hypolimnion. This facilitates the deposition of the particles retained along the thermocline which consequently tends to fade away. Melting of ice at the end of winter leads to a general cooling, until the thermocline is established again as a consequence of spring warm-up.

Density currents in lakes are caused by fluvial discharge (Fig. 5.11). The surficial flux or overflow leads to plumes of warm, lighter water brought in by the summer circulation. They result in laminated sedimentary deposits in which the grain size decreases away from the shore, and which are frequently gray to black in color due to a high content of organic matter. The intermediate flux or interflow is made up of waters with densities intermediate between those of the epilimnion and the hypolimnion which spread out below the thermocline with their suspended matter. These particles settle to the bottom after the inversion of densities during autumn has led to the disappearance of the thermocline. They form the lighter-colored laminae of the cold season. The deeper flows or underflows result from high-energy turbidity currents or from the intrusion of dense cold waters following the snow melt in spring. These flows introduce oxygenated waters rich in coarser particles to the lake bottom.

5.3.1.3 Chemical Composition

The chemical composition of lacustrine waters depends mainly on the relation between the supply of dissolved matter and the rate of evaporation. In cold or temperate lakes, evaporation is moderate, and the oversaturation required for the precipitation of low-Mg calcite will only occur in summer when phytoplankton produces large quantities of CO_2 as in Lake Zurich (Kelts and Hsü 1978). When organic production is high, the environment can turn eutrophic, leading to a considerable oxygen deficiency in water layers not stirred by winds: the resulting sediments become reducing. In the opposite case, the water column remains oxygenated: the lake then is oligotrophic and its sediments are light-colored. In lakes of arid regions, evaporation mostly exceeds the rate of influx, and saline lakes with above 5000 ppm dissolved matter will result. The establishment of such lakes is favored by the absence of a discharge outlet (closed basins) as in the case of the Dead Sea, and by rain on the basin slopes, but not above the lakes themselves. Other supporting factors are drainage from older evaporitic sequences in the source area like the streams feeding the Dead Sea, and the presence of alluvial fans around the lake which trap the particulate matter. Numerous saline lakes are ephemeral like continental sabkhas on playas (cf. Sect. 5.2.2), or their areal extent is subject to seasonal fluctuations, i.e., shrinkage during the dry season.

5.3.2 Recent Lacustrine Deposits

5.3.2.1 Terrigenous Supply

In general, the sediments deposited along the edge of lakes are rather heterogeneous, being subjected to fluvial processes, coastal erosion, and reworking, as well as to chemical precipitation. In contrast to this, the deeper lacustrine sediments are more regular, as they are intensely influenced by stratification in the overlying body of water. In lakes dominated by detrital supply, the rate of sedimentation is frequently up to ten times that of the marine environments (Galloway and Hobday 1983). In Lake Brienz of Switzerland, gravitational phenomena and exceptional flood discharges take place once or twice every century, leading to erosion of the delta front and the formation of slumps with vertical and horizontal size grading in resulting turbidites (Sturm and Matter 1978). The more frequent underflows lead to weak seasonal turbid currents which deposit silty-clayey laminae throughout a large part of the basin (Fig. 5.11). These fine-grained dark layers resemble very much those deposited by settling from summer overflows of identical provenance which alternate with the lighter-colored winter deposits formed after the disappearance of the thermocline (interflows, Sect. 5.3.1.2). Farther removed from the points of entry of fluvial discharge, fine-grained muds are deposited which are homogeneous above the thermocline and laminated below. In contrast to this, at the point of entry of considerable fluvial supply, for example where the Rhone enters Lake Geneva, a sublacustrine delta with a rather complex detrital fan tends to develop. It is characterized by channels, detrital levels, lateral migration, and gravitational sliding with numerous sedimentary structures such as current ripples, sand waves, cross-bedding, convolute bedding, etc. (Houbolt and Jonker 1968). Water level fluctuations are much more frequent in lakes than along the margins of the sea and lead to particularly complex sedimentary successions.

5.3.2.2 Organic Content

A high content of organic matter in the sediments results either from up-river retention of detrital material by alluvial cones or barrages as in Lake Zurich, or from a particularly high productivity of phytoplankton (chrysophyceae, diatoms) in generally productive regions as the East African Lake Kion (15% C_{org}) or in lakes rich in silica which occur in volcanic craters or in higher portions of regions underlain by siliceous rocks. Seasonal rhythmites or varves are developed in the deeper portions of lakes, whereas along the edges they are frequently destroyed by storms or sublacustrine slumping.

5.3.2.3 Chemical Influences

In lakes with pronounced chemical supply, *evaporites* may develop in a concentric pattern with the most soluble salts being precipitated last in the central portions. In the playas of Saline Valley in California, halite crystallizes in the centers of depressions filled by highly concentrated brines after glauberite ($CaNa_2(SO_4)_2$), gypsum, calcareous tufa and travertine, and even algal crusts formed around the periphery (Fig. 5.12). Along the edges, detrital formations accumulate which are derived from ephemeral water courses and sometimes reworked into dunes by wind action. Alternation of detrital and evaporitic layers in saline lakes usually results from cycles involving periods of strong precipitation and intense evaporation. The results are bedded halite deposits in which vertically oriented halite crystals are arranged in layers separated by bands of terrigenous material deposited by water or even aeolian activity.

A variety of minerals characteristic of lacustrine environments tends to form in *volcanic or hydrothermal environments* as in certain basins along the East African rift valleys. This is especially the case for trona (Na_2CO_3, $NaOHCO_3$, $2H_2O$) forming a 40–60-cm thick layer in Lake Magadi of Kenya together with authigenic zeolites and complex salts such as erionite ($NaKAl_2Si_7O_{18}$, $6H_2O$), magadiite ($NaSi_7O_{13}(OH)_3$, $3H_2O$), kenyaite ($NaSi_{11}O_{20,5}(OH)_4$, $3H_2O$) (Surdam and Eugster 1976). In Lake Bogoria of Kenya, the sediments deposited 20000 years ago are primarily alkaline and anoxic, with analcime, mordenite, fluorite, gaylussite, siderite, and eventually material derived from freshwater or sodium-carbonate waters (trona, kanemite $NaHSi_2O_4(OH)_2$, $2H_2O$, nahcolite $NaHCO_3$, magadiite, etc.), depending on whether the water level of the lake was high or low (Tiercelin et al. 1982). In Lake Albert and Lake Manyara, Singer and Stoffers (1980) described the illitization of smectites by potassium derived from the transformation of Na-K zeolites into analcime.

In certain lakes *other types of clay minerals* are also formed. The presence of highly reactive silica from diatoms can give rise to nuclei leading to the crystallization of smectites in saline environments such as the Bolivian

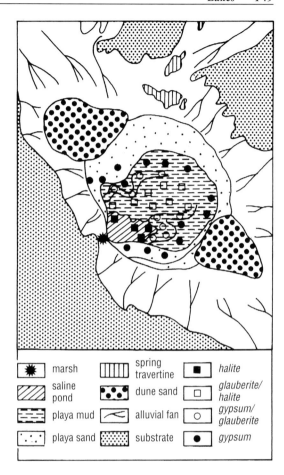

Fig. 5.12. Sedimentary facies and evaporitic deposits in the playa of Saline Valley/California (after Hardie et al. 1978)

Altiplano (Badaut 1981). Strong evaporation in a closed basin leads to the concentration of dissolved matter derived from a region of intense chemical weathering, as in Lake Chad which during the Pleistocene shrank from a surface area of 300000 km² to about 100000 km² (Fig. 5.13). The argillaceous deposits formed on the resulting plain consist of detrital minerals such as gibbsite, goethite, illite, and Al-Fe beidellite supplied by the Chari River, as well as of carbonates, various salts (gaylussite, trona), and amorphous silica formed in the evaporitic marginal zones. These lead to granular nontronite developing in the delta during dry periods, to Al-Fe smectite (beidellite) in the vertisols of the flood plains and partly redeposited in the central sub-basin, and to authigenic stevensite in the intermediate arms of the northern sub-basin (Carmouze et al. 1977). Only the development of minerals not present in the respective source areas will allow the ready distinction between processes of authigenesis and detrital supply, reworking of soils along the margins, and of mechanisms of differential sedimentation.

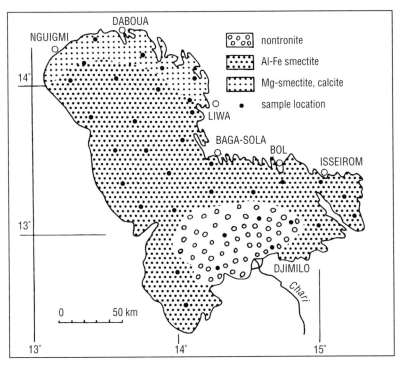

Fig. 5.13. Distribution of smectites in sediments of Lake Chad (after Carmouze et al. 1977)

5.3.3 Ancient Lacustrine Deposits

The identification of lacustrine environments in geological sequences is often difficult, especially as the deposits of ephemeral evaporitic lakes tend to be rather similar to those of littoral marine sabkhas. The prime distinguishing feature is the absence of marine flora and fauna in the lacustrine deposits. This is of little use in Precambrian sequences which are mostly devoid of fossils, and where the depositional environments are poorly known. Furthermore, lacustrine series of pronounced chemogenic nature are generally azoic, containing sometimes minerals such as trona, borates and complex salts (cf. Sect. 5.3.2) which are not able to precipitate in marine environments. Sedimentary structures are also sometimes helpful in identifying lacustrine deposits. Structures such as herring-bone cross-stratification and flaser bedding resulting from tidal currents are absent, whereas hummocky structures of low wavelength (40–80 cm), varved rhythmites and upward-coarsening or emersive sequences resulting from the filling of lakes may be frequently observed (Fig. 5.14).

Mixed carbonatic and siliciclastic lacustrine sedimentation occurred to about 4000 m thickness in the Caithness Flagstone Group of the Cread basin

Lakes   151

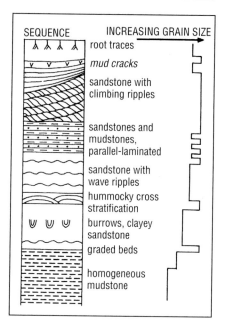

Fig. 5.14. Idealized sedimentary succession in a lacustrine environment, after data from the Permo-Triassic of the eastern Karoo, South Africa (after van Dyke et al., in Galloway and Hobday 1983)

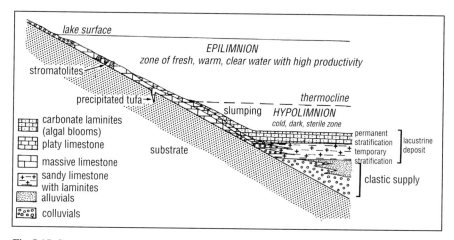

Fig. 5.15. Sequences and sedimentary environments along the edge of the mid-Devonian Orcad basin of northeastern Scotland (after Donovan 1975)

of northeastern Scotland during the mid-Devonian. These deposits, forming part of the Old Red Sandstone, are intercalated between fluvial sandstones of lower and upper Devonian age. They abound with laminites particularly rich in silty-clayey layers showing subaqueous shrinkage and subaerial desiccation structures, followed by carbonate-rich material resulting from seasonal algal precipitation (Fig. 5.15). The sedimentary succession observed along the edge of the basin results from fluvial sand bodies with decreasing grain size,

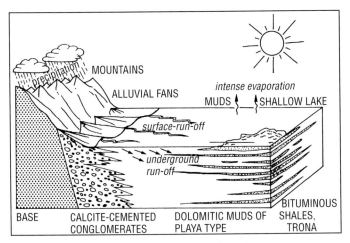

Fig. 5.16. Main sedimentary environments of the Eocene lacustrine Green River Formation (after studies in the Wilkins Peak Member, Wyoming/USA by Eugster and Hardie 1975)

alternating with seasonal deposits derived from the stratification of the lake water (in Donovan 1975). This non-evaporitic, rapidly subsiding intramontane basin, the water cover of which probably was permanently stratified, was filled within about 10 million years.

Predominantly chemical and organic sedimentation characterizes the Eocene Green River Formation of Wyoming/USA. This about 950-m thick succession entails a number of sedimentary facies ranging from marginal detrital alluvial fans to the central playas. Evaporitic conditions in an arid climate are marked by the development of a huge saline lake and by marginal oolite shoals, algal crusts, oncolites, and reef structures (Picard and High 1981). Within the Wilkins Peak Member in Wyoming, six facies can be distinguished in the Eocene paleogeography of the region (Fig. 5.16; Eugster and Hardie 1975):

1. Conglomerates with dolomitic mudstone clasts, probably the result of reworking of algal crusts and massive concentrations along the margins of the evaporite basins.

2. Sandstones and calcareous mudstones with wave ripples and cross-laminations formed close to shore.

3. Mudstones with mud cracks and siltstone laminae, deposited in the central playa, the extent of which repeatedly changed with time.

4. Organo-dolomitic laminites and breccias of bituminous clays rich in hydrocarbons. Representing immense potential oil reserves, these deposits result from gelatinous algal and bacterial oozes which accumulated on a subsiding, but intermittently emergent, lake bed.

5. Dolomitic mudstones rich in sodium carbonates which represent the largest recent reserves of trona, resulting from increased chemical concentration of the playa brines.

6. Immature sandstones with cross-bedding and channeling, derived from sands deposited in detrital alluvial fans at the foot of mountain ranges surrounding the lacustrine basin.

5.4 Alluvial Fans

5.4.1 Introduction

Alluvial fans are cone-, lobe-shaped, or semicircular accumulations of sediment, deposited at the *foot of mountain ranges* (Fig. 5.17). These piedmont deposits, usually of rather small extent, are dominated by coarse detrital sediments left behind by strong flows of water discharged from mountain ranges, along the side of mountain valleys, or at the foot of glaciers (cf. Heward 1978; Galloway and Hobday 1983). They occur in the upper reaches of fluvial discharge basins, or spread as subaqueous fan deltas directly into lakes and even marine environments subjected to wave or tidal action. Alluvial fans are mainly found in subarid environments, although extensive deposits occur in humid climates over faint slopes, as the Kosi River fan discharging into the River Ganges from the foothills of the Himalayas over an area of 150 × 120 km. They result mostly from intense erosion in mountain ranges subjected to

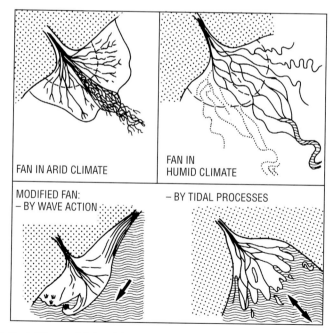

Fig. 5.17. Main types of alluvial fans and deltas (after Galloway and Hobday 1983)

strong tectonic activity, although they can also be caused by climatic changes, the melting of high-altitude glaciers, lowering of the erosive base level, or the destruction of plant cover. Alluvial fans tend to be better preserved when they are deposited along fault zones associated with strong subsidence. They may contain economic concentrations of minerals, such as gold and uranium in the late Archaean Witwatersrand Supergroup of South Africa, or also coals and hydrocarbons.

The main mechanisms responsible for sedimentation in alluvial fans are *torrential discharges,* especially in humid climates, and poorly sorted *debris flows* which predominate in drier climates. With these mechanisms are associated a number of other processes such as aeolian action which redistributes sandy-silty material in zones with sparse vegetal cover, binding of fine-grained particles within masses of gravel and boulders. These masses act as a screen when the water infiltrates downwards after floods. In arid zones, the surface tapping of phreatic horizons leads to the precipitation of gypsum, calcite, and carbonate crusts or nodules, whereas in humid zones swamps and bogs develop. Lastly, the action of waves or tides shapes the distal portion of fans debouching into water-filled basins (Fig. 5.17).

5.4.2 Alluvial Fans in Humid Regions

Modern fans of various sizes have been described in detail from Nepal, India, Alaska, Honduras, and Iceland (Galloway and Hobday 1983). Active mainly during the rainy season or the snow-melt, these fans are characterized by continuous fluvial discharge, slight slopes, and rates of sedimentation which fluctuate with time, between wide limits (up to 7 m per year). Other features are debris flows in the proximal portions and the development of vegetal cover over inactive parts. The mean size of constituent particles decreases in the downstream direction (Fig. 5.18). In the upstream section, boulders and pebbles are transversally arranged and frequently rounded, showing rough horizontal bedding in individual bars, and cross-beds, the concave side of which points downstream in channels shaped by torrential discharge. Incidence of cross-bedding increases with the number of channels in the fan midsection, whereas the bars grade into sheet flood deposits of rhombohedral form, becoming more drawn out as the flow dissipates.

Alluvial fans of the humid type are known from a large number of *fossil environments.* Their abundance in pre-Carboniferous sequences is probably linked to sparse vegetal cover which favored intense erosion of mountainous regions (Schumm 1977). Some of these fans were rather large, like those of the

Fig. 5.19 A, B. Reconstruction in plan view (**A**) and section (**B**) of the humid alluvial fan in the Cambrian Van Horn Sandstone, Texas (after McGowen and Groat 1971). The *arrows* indicate the sense of sediment dispersal

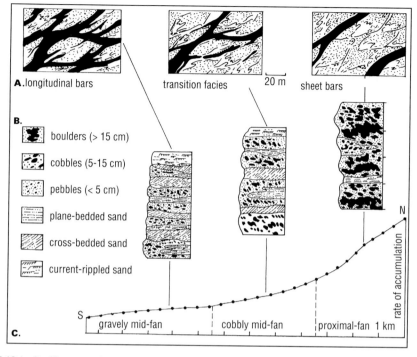

Fig. 5.18 A–C. Changes of sedimentary sequence in an alluvial fan of humid climates in Alaska (Boothroyd 1972, in Reading 1986). **A** Interchannel deposits; **B** sequence of sediments; **C** rate of accumulation (on arbitrary scale)

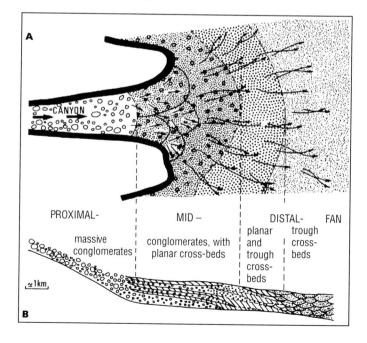

Cambrian Van Horn sandstone of Texas (McGowen and Groat 1971) which show pronounced size grading and a succession of characteristic sedimentary structures in the down-fan direction (Fig. 5.19). Proximal facies are characterized by thick boulder beds deposited by strong gravity flows passing laterally and upwards into sheets of sand bars. The distal portions show an increase of sandy and even silty-clayey deposits, and the planar and concave cross-beds of anastomosing channels.

5.4.3 Alluvial Fans in Arid Regions

Fans developed in semi-arid regions are usually small and conical, forming along the edges of continental rift zones such as the Dead Sea, the Persian Gulf, and California. Marked by an abundance of debris flows and deposits of surficial sheet floods across virtually the whole extent of the fan, they contain few channels resulting from torrential discharge, and infiltration deposits are restricted to the proximal portions (Fig. 5.20). The various grain-size fractions are distributed uniformly and without sorting, horizontally and vertically throughout the alluvial deposits, with the exception of large boulders that are not as easily transported. The debris flows show little clear stratification, and the coarser fragments engulfed in the heterogeneous matrix are usually angular.

Fossil alluvial fans, like their modern counterparts, are characteristically quite small, containing poorly sorted sediments of compositional and textural

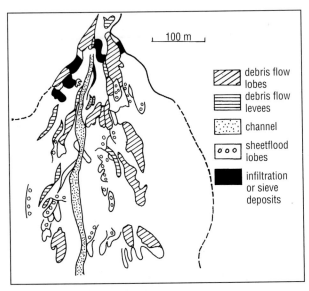

Fig. 5.20. Sedimentary facies in an arid alluvial fan in Death Valley, California (after R. L. Hooke, in Galloway and Hobday 1983)

immaturity with few current structures. Typical examples are found in the Permo-Triassic of Scotland and Spain. The aridity of the depositional environment may be indicated by lack of vegetal cover, and by paleosols with calcareous crusts of the caliche type, sandy aeolian intercalations, and by evaporitic playa deposits in depressions in the more distal portions of the alluvial cones (Galloway and Hobday 1983).

5.5 Rivers and Streams

5.5.1 Introduction

Rivers are the prime agents responsible for collecting detrital particles derived from weathering, and for transporting them to the receiving marine or lacustrine basins (cf. Schumm 1977; Miall 1978; Walker and Cant, in Walker 1984). The lateral extent of fluvial deposits is usually small, save for alluvial plains developed over peneplained shields, and in certain coastal plains. The fossilization of fluvial deposits is favored by zones of strong subsidence and accumulation like those developed along expanding oceanic basins or intramontane discharge basins bordered by fault zones. The abundance of sandy facies, frequent lithological diversity, and the presence of accumulations of organic matter, predispose fluvial deposits for trapping hydrocarbons and for the formation of uranium deposits and coal. Fluvial morphology is dominated by the development of *channels* for water discharge. Their shape varies from rectilinear in areas of moderate flow velocity and with fine-grained fill, to the anastomosing unstable types with rapid flow resulting in longitudinal bars and coarse-grained deposits. Intermediate types are made up of fluvial meanders of varying sinuosity marked by prograding bars along the convex banks and the appearance of longitudinal bars (Fig. 5.21).

The fluvial depositional mechanisms in channeled zones depend on velocity and turbulence of the water masses. Maximum velocity and turbulence are encountered under very shallow cover of water and on the concave banks where erosion is greatest. Deposition occurs in troughs or pools, and on the convex bank where flow is slower and more laminar. Turbulent flows lead to minor fluvial deposits when cutting across from one meander to the next in the downstream direction. The sinusoidal course of the river leads to a second-order vertical helicoidal movement at right angles to the flow of water resulting in sedimentation on the inner side of the meanders (Fig. 5.21). The respective deposits are characterized by a *lateral progression* of sedimentary units (Fig. 5.22).

During floods, the more rapid currents tend to attack the banks of the meanders where these are no longer in line with the flow contours. This results in a cutting of the meander and a failure of the bank, leading to flooding along the major river courses. Deposits thus formed show a vertical aggradation of

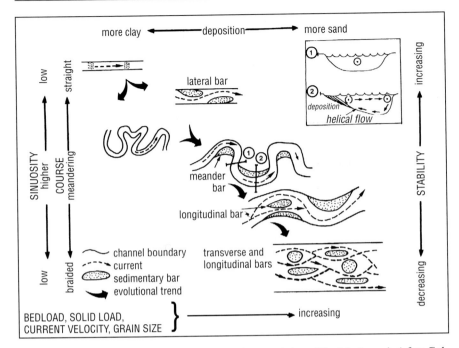

Fig. 5.21. Morphology and hydrosedimentary characteristics of fluvial channels (after Galloway and Hobday 1983)

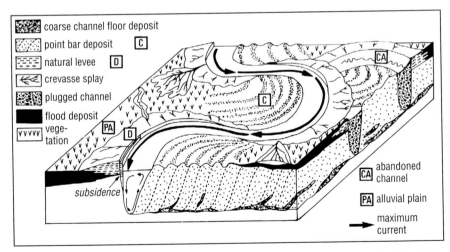

Fig. 5.22. Main mechanisms and types of sediments in the valley of a meandering river (after J. R. L. Allen, in Fairbridge and Bourgeois 1978). Note: vertical scale exaggerated

sedimentary units on the alluvial plain. Locally, the flood currents can avail themselves of old channels or depressions where they lead to breaching of the margins and the formation of crevasse splays and bank failures. The flow then tends to diminish, allowing the deposition of material from suspension with the grain size decreasing along the axis of the current.

Another important mechanism is the abandoning of main and secondary channels. Flood deposits along the margins of migrating meanders lead to a gradual rise of the banks and thereby to the establishment of a channel raised above the main course of the river. The focal failure of the banks during an exceptional flood facilitates the emptying of the channel into the surrounding alluvial plain where it then will establish a new course farther downstream. Other channels may be abandoned as a result of the breach of the convex side of migrating meanders by turbulent currents. Abandoned channels are characterized by coarse sediments frequently overlain by fine-grained flood deposits and even by soil horizons (Fig. 5.22).

Altogether six different types of sedimentary facies occur in fluvial environments (Happ et al., in Blatt et al. 1980):

1. point bars or deposits laterally prograding along the convex banks of a channel;

2. lag deposits or coarse-grained material deposited by violent currents on the channel bed;

3. fine- to coarser-grained deposits, sometimes rich in organic material, resulting from the filling of abandoned channels;

4. deposits accreted vertically on the alluvial plain over fairly large areas with well-developed bedding;

5. overflow deposits resulting from isolated breaches of the banks during floods, of localized extent with grain size sometimes decreasing upwards and along the course of the flow;

6. deposits resulting from slumps along the edges of the valley and reworked together with flood deposits.

5.5.2 Sedimentary Sequences in Recent Fluvial Deposits

Channels with low sinuosity, or developed braiding, formed in fluvial environments of high energy are predominantly sandy- or clayey-sandy.

1. Sandy sediments are formed in bars along the banks, or elongated along the bed of the channel itself, migrating downstream during floods in little inclined beds with planar structures. The base of these sedimentary sequences is plane, despite scouring, whereas the sediments themselves are coarse, showing concave channel cross-beds. The tops of the beds are marked by soils and plant traces as a result of emersion of the bars (Fig. 5.23 A).

2. Sandy sedimentation also occurs in bars oriented at right angle to the current. They are particularly frequent in rapidly flowing, braided fluvial

160   Continental Sedimentation

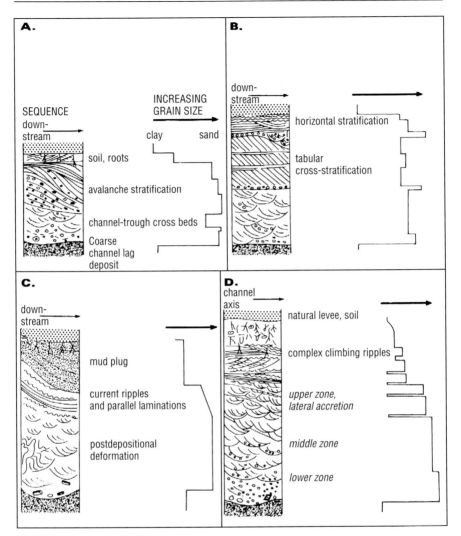

Fig. 5.23 A–D. Idealized sedimentary sequences developed along the course of a low-sinuosity river (after Galloway and Hobday 1983). **A** Longitudinal sand bar; **B** transverse sand bar; **C** sandy-argillaceous fill in a braided system; **D** for comparison, sequence on the convex bank of a high-sinuosity river

systems. During floods, the particles carried along the bottom are deposited on the downstream side of the bars in parallel layers inclined at fairly high angles, above which the normal horizontal deposits advance. Transverse bars are highly mobile and rarely emerge above the water level (Fig. 5.23 B).

3. The less common, fine-grained material occurs in channels with concave bottoms which tend to become more symmetrical as their course decreases in sinuosity. The sandy-clayey sediments are deposited by slow cur-

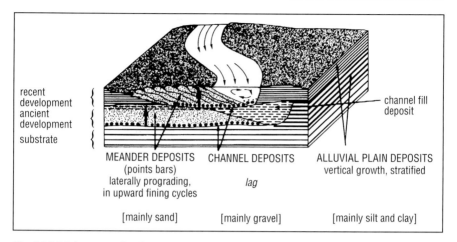

Fig. 5.24. Main types of sediments along a high-sinuosity meandering river (after Blatt et al. 1980)

rents in channels removed from the main course of the river and frequently show structures such as compaction or load casts resulting from post-depositional deformation. Slight upward decrease in grain size occurs, together with small current ripples and laminations, and layers of clayey material (Fig. 5.23 C).

Highly sinuous meandering channels exhibit features indicative of convex and concave river banks (Fig. 5.24):

1. On the convex side, deposition progresses laterally towards the channel axis in sequences showing upward decrease in grain size resulting from diminishing currents, and with cross-beds inclined against each other. Each sequence results from a flood or another significant variation in flow regime. They also are formed above coarser deposits on the channel bottom, showing inclined concave beds and cross-stratification. They are terminated by current ripples, climbing ripples, planar beds, and laminations indicative of thin water cover and even emersion (Fig. 5.23 D). Each sequence shows a surface ridge separated from that of the preceding unit by a depression. The units can eventually be covered by horizontal flood deposits.

2. Size and shape of the concave bank are well maintained as it migrates to the outside of the meander, thereby ever increasing in curvature. During flood events, the flow of waters across the banks leads to an abrupt reduction of the flow velocity and to rapid deposition of material on the banks of the channel as horizontal beds, establishing a natural dyke.

3. On the bottom of channels of all types, coarser particles accumulate in the form of a lag deposit. They are again subjected to erosion during the next incident of increased velocity and turbulence. Subaqueous sand waves migrating downstream usually cover these scattered coarser deposits. They exhibit numerous concave-upward cross-beds, the size of which is controlled by the energy of the currents.

The main course of a river and the emerged inter-channel areas covered by water during floods are places of irregular and usually moderately fine-grained horizontal sedimentation. These deposits which tend to be better preserved in zones of permanent flow, may become reworked by a migrating channel or as a consequence of abandonment of an active channel. Usually an overall decrease in grain size occurs in the direction of the borders of the alluvial plain where the fluvial deposits are in contact with coarser material directly derived from the source areas. In dry regions where the groundwater level is fairly deep, the alluvial plains may be subjected to temporary aeolian deflation. In humid regions, the groundwater level only appears on the surface in bogs, swamps, ponds, and lakes.

5.5.3 Ancient Fluvial Environments

The reconstruction of ancient fluvial environments is based on fairly recent examples. Quaternary environments are controlled by high and low stages of the sea-level as a result of cold glacial and warmer interglacial intervals. These lead to the formation of terraces by repeated incision of meandering rivers into alluvial formations or into their solid substrate. During stages of high sea-level, alluvial terraces form at progressively deeper levels as the erosion cycles succeed each other. Size of sedimentary particles tends to decrease as a valley becomes filled during a given cycle, because the energy of the waters and the erosive force diminish when base level rises and the continental ice-caps disappear.

Thus, during the Holocene climatic warming, the Mississippi valley became filled by a succession of sediments starting, on a Tertiary base incised by upper Pleistocene torrents, with coarse-grained sediments supplied by a system or anastomosing channels. This was succeeded by fine-grained sandy-silty layers deposited in a system of sinuous meanders with marginal swamps (Fig. 5.25).

Fluvial successions identified in ancient series frequently exhibit an upward decrease in grain size (fining-upward) and a reduction in the diversity of sedimentary structures as a result of mechanisms such as lateral migration of meanders, plugging of channels, flood deposits covering the alluvial plain, and rise of the base level. The 400-m thick fluvial deposits of the lower Devonian Old Red Sandstone of southern England consist of a number of 3–18-m thick sequences or cyclothems. Each cyclothem fines upward from basal erosive contacts with flute casts and conglomeratic lag deposits on the bottom of the channels, to current ripples and erosive structures, siltstones deposited over the flood plains, and eventually to pedogenic carbonate crusts (Fig. 5.26). These frequent indicators of a drop in energy, sometimes terminated by emersion leading to the formation of soil horizons and mud cracks, are the result of preferred fossilization of channel fill sediments. During phases of increasing

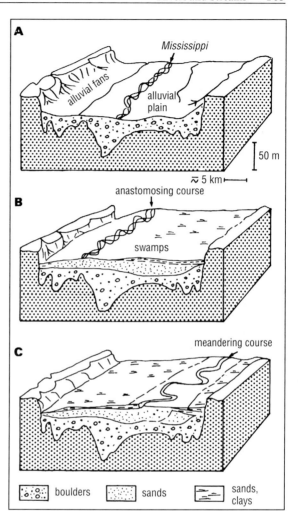

Fig. 5.25 A–C. Stages of filling of the Holocene Mississippi Valley, after incision into the Tertiary base to the lower marine sea-level (−130 m) of the Pleistocene. **A** Gravel and coarse sand deposited by a high-energy braided fluvial system, migrating laterally and bordered by alluvial fans (sea level at −30 m); **B** sand deposited up to a sea-level of −6 m in the Gulf of Mexico, with levees permitting the establishment of swamps along the course of the river; **C** recent sand, silt, and clay deposited by a medium-energy meandering river raised above the alluvial plain (after Fisk 1944)

fluvial energy, the sedimentary material is remobilized and transported downstream.

The succession of fluvial facies along the course of an ancient river typically shows a decrease in grain size with a flattening of slopes, widening of the valleys, and a drop in the erosive force of the waters in the lower reaches of the river system. The Devonian Hornelen Basin of Norway is an excellent fossil example of this zonation (Steel and Aasheim 1978). It was a strongly subsiding basin bordered by subvertical faults, in which the conglomeratic upstream sediments are followed in the downflow direction by sandy deposits from anastomosing channels, then by distal clayey silts supplied by floods, and eventually by lacustrine deposits (Fig. 5.27). Along the edge of the main basin, secondary alluvial fans were built up by torrential flows or debris flows (cf.

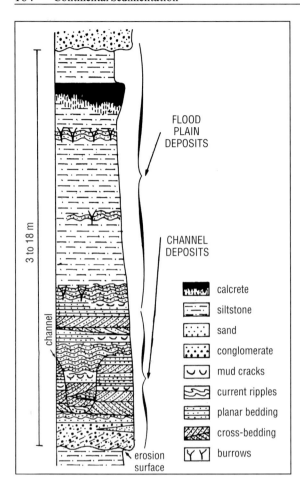

Fig. 5.26. Typical cyclothem showing lower Devonian fluvial sequence of Galicia/Spain (after Allen and Tarlo, in Gall 1976)

Fig. 5.27. Sedimentary successions deposited from upstream to downstream areas of the Devonian fluvial Hornelen Basin (Norway). The *inset* illustrates the increase in grain size resulting from progradation of the detrital complex with time (after Steel and Aasheim 1978). The *arrows* indicate the sense of dispersal

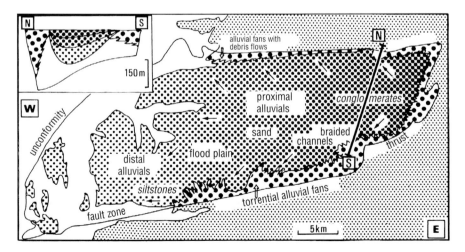

Sect. 5.4). With time, the repeated subsidence of the basin led to innumerable renewals of erosion facilitating active sedimentation leading to a prograding advance of the coarse proximal facies over finer-grained distal sediments. Such a downstream progradation in an unstable tectonic environment leads to a coarsening-upward succession gradually filling the fluvial depository.

6 Marine Coastal Environments

6.1 Deltas and Estuaries

6.1.1 Hydrodynamic and Sedimentational Mechanisms

6.1.1.1 Deltas and Estuaries Represent Transitional Hydrodynamic and Sedimentary Environments, Exposed to Fluvial and Marine, and Sometimes Lacustrine Influences

In *deltas,* fluvial influences predominate and lead to an irregular advance of the coastline into the open sea. Processes of sediment accumulation exceed those of reworking or destruction by wave or tides. The classic delta consists of subhorizontal top-set beds of the delta plain which is partly emerged and dissected by channels, and deposited above the more inclined foreset beds of the delta front. These, in turn, grade into subaqueous bottom-set beds of the pro-delta. In submarine environments, deltas frequently grade into deep-sea detrital fans (cf. Sect. 7.1). The phenomena of progradation and of lateral migration of deltaic channels with time lead to considerable rates of accumulation of sediments of highly variable grain size, covered and buried as the result of strong subsidence. Thus, deltas can form important reservoirs of oil, coal, uranium, etc. formed in reducing environments (Broussard 1975; Coleman 1981; Blanc 1982; Galloway and Hobday 1983; Elliott, in Reading 1986).

In *estuaries,* marine influences and especially those of tidal processes are important. A typical estuary represents part of a river invaded by strong tides (salt water and freshwater pushed back by high tides), in which the salinity decreases in the upstream direction. The antagonism between fluvial and marine processes expresses itself in important differences in salinity and movement of water along the direction of flow and from the center of the estuary to the sides. Despite the dynamic influences exerted by floods and tidal currents, estuarine deposits are characterized by an abundance of silty-clayey material, especially in tidal mud flats along the main channels, in grass-covered beach zones inundated only during exceptionally high tides, and in the adjacent lagoons and tidal marshes (cf. Lauff 1967; Wiley 1976).

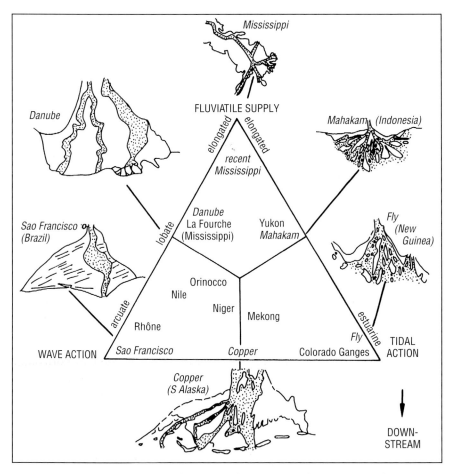

Fig. 6.1. Classification of deltas as a function of the relative importance of fluvial and marine processes (after Galloway and Hobday 1983)

Deltas and estuaries basically represent a continuous series of sedimentary environments at the interface of alluvial plains and basins in which fluvial processes interact with, and counteract the influences of tides and waves (Fig. 6.1). In such a system, an estuary represents an extreme type of a delta subjected to tidal influences. If, at any stage of its evolution, an estuary is dominated by fluvial supply, it tends to become plugged from upstream and to grade into a seaward advancing delta. This applies to the Gironde, the estuary of the River Garonne of southwestern France which under the present natural conditions will become filled within about 1000 years (Castaing 1981). Conversely, a delta in which tidal processes gradually replace fluvial influences, will evolve into a typical estuary. This was the case for a large number of rivers originally debouching into macro-tidal seas. During the gradual rise of the sea-level resulting from the last Quaternary deglaciation, the lower reaches of their alluvial valleys were invaded again by the sea.

6.1.1.2 The Antagonism Between Marine and Fluvial Influences, So Notable in Estuaries, Manifests Itself in a Number of Ways

The *tidal wave* which travels in a sinusoidal manner in the open ocean is deformed in the estuarine environment by friction between the water and the river bed and by the upstream narrowing of the banks (Salomon and Allen 1983). The bottom friction causes a stronger retardation of flow along the bottom than in the higher parts of the water column, and leads to a wave of increasing asymmetry upstream. This results in a shorter duration of the flood than of the ebb stage and to stronger currents during the former. The narrowing of the estuarine cross-section is accompanied by a notable upstream increase of the tidal amplitude. Commonly, the effects of the narrowing cross-section exceed those of the bottom friction, as in hypersynchronous estuaries like the Gironde. Where the frictional forces predominate, as in estuaries of shallow depth and relatively constant width, the amplitude of the tidal wave decreases gradually in the downstream direction. This is the case in the hyposynchronous Picardie estuaries along the French coast of the English Channel. Estuaries in which the influence of friction and narrowing are roughly equal are referred to as synchronous.

The mixing of marine- and freshwaters leads to rather complex horizontal and vertical gradients in salinity, controlled mainly by the rate of fluvial discharge, the tidal differences, seasonal trends, and by the morphology of the

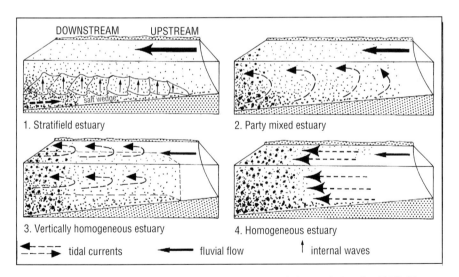

Fig. 6.2. Types of estuarine circulation (after Pritchard and Carter, in Leeder 1982). For explanation refer to text. The *points* represent the suspended particles

estuary in question. Based on these factors, Pritchard and Carter (in Leeder 1982) distinguish four types of estuaries (Fig. 6.2):

1. Stratified: Here the influence of tides and surf is weak, permitting the establishment of a discontinuity between the denser sea-water and the overlying freshwater. The former persists as a marked saline wedge over varying portions of the estuaries, depending on the rate of fluvial discharge and, to a lesser degree on the rise of the water level. Vertical exchange of these waters is weak. Sediment load is carried out to the sea mainly during floods.

2. Partly mixed: Here the turbulence resulting from the tidal current destroys the interface between salt and freshwater, leading to vertical exchanges. A vertical gradient in salinity is established and the transport of particulate matter is subjected to tidal as well as fluvial influences.

3. Vertically homogeneous: Here an intense mixing of fresh- and saltwater is affected by the tidal currents. There is only a horizontal seaward increase in salinity, and a lateral gradient of the water level caused by the Coriolis force resulting from the eastward rotation of the earth. In the northern hemisphere, the saline layer is thicker and flows more rapidly along the left bank of an estuary. Particulate load is partly of marine origin, and distribution of the sediments is mainly controlled by tidal currents.

4. Vertically and laterally homogeneous: Here marine influence affects only the sedimentation. This is the case in narrow bays with little fluvial sedimentary discharge, in which there may exist a moderate horizontal salinity gradient.

The sequence from type 1 to 4 represents a continuous passage from extremely deltaic to estuarine conditions. In a given estuary, depending on the time of the year, one or more, or all types of estuaries may coexist (Allen et al. 1981).

The antagonistic and complementary movements of continental discharge and marine waters exert a number of effects on the transport of particulate matter in estuaries. The turbidity of estuarine waters may considerably exceed that of upstream continental and downstream marine waters. In the case of the Gironde, the concentration of suspended matter is between 0.1–1 g/l at Libourne in the upper reaches where it is referred to as the Garonne, and several dozens of grams/liter in the middle portion of the estuary near its mouth at Grave (Allen et al. 1981). The suspended matter travels with the tidal cycles which permit the distinction of different characteristics (e.g., Despeyroux 1985; Fig. 6.3). Turbid waters with more than 1 g/l suspended matter frequently form a *clay plug* which, according to recent studies, establishes itself under the influence of density gradients resulting from changes in salinity or by purely mechanical tidal processes (cf. Castaing 1981). The clay plug usually migrates up the estuary during high tide and seaward during low tide, and can be expelled by exceptionally low tides or fluvial floods to contribute to marine sedimentation in the open sea. In the Gironde, this clay plug represents a mass of more than 2 million tons, resulting essentially from retention in suspension, and from blockage of the fluvial discharge, by tidal dynamics (Allen et al. 1981; Castaing 1981). Clay plugs are absent

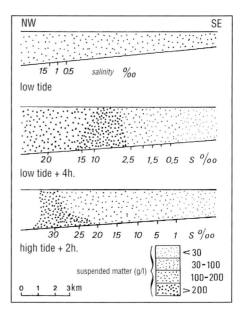

Fig. 6.3. Movement of suspended matter during a neap tide cycle in the Canche estuary, French coast of eastern English Channel (after Despeyroux 1985). Tidal coefficient 111, mean fluvial discharge 10.8 m³/s

from, or only intermittently present in, stretched estuaries with high discharge rates like the Picardie coast of northwestern France. Individual lenses with elevated concentrations of suspended matter (up to 400 g/l in the Gironde) forming close to the bottom are referred to as *fluid mud* (Allen et al. 1981). Such dense suspensions accumulate in the tidal channels during the change of tides. They constitute important reservoirs of particulate matter (2–3 million tons in the Gironde), contributing to the clay plugs.

Deposition of mud in estuaries occurs in several ways: coagulation and flocculation of negatively charged clay particles coming into contact with saline waters rich in cations, formation of organo-mineral aggregates, settling between turbulent zones above tidal channels, accretion of clay particles to the clayey bottom sediments during the retreat of the sea at low tides. The phenomena of flocculation frequently appear to be of secondary importance compared to those of accretion of particles (Chamley 1974; Allen et al. 1981). Granulometric sorting also occurs in association with mineralogical sorting. Clay minerals in particular are frequently subjected to important modifications during their passage from fresh into salt waters, leading to *differential settling* and to ionic exchange on the surfaces of the clay sheets. The smectites which are smaller than most other clay minerals and show a lower tendency to flocculation and better floatability (e.g., Gibbs 1977), will settle out less easily in estuaries subjected to strong currents than upstream in the river or in the basins seaward of the estuary. This is shown by the relative increase of smectite in suspensions compared to estuarine sediments, such as in the Loire (Manickam et al. 1985), by the complete reconstitution of the clay mineral spectrum above and below the estuary, as in the Guadalquivir of Spain

(Mélières 1973), and by the constancy of geographic and chronological distribution (e.g., the Guadalupe Formation of Texas; Morton 1972). During their passage through an estuary the dissolved chemical elements show rather varied behavior. In the Gironde, the rare earths, alkalis, alkaline earth as well as Fe and Sc do not vary much, whereas Zn, Cu, Ni and Pb show a minimum in the downstream direction increasing again near the mouth (Jouanneau 1982). These variations are independent of the salinity, the content of particulate matter, and the type of clay minerals. They appear to be tied to fluctuations in the concentration of particulate organic carbon, and controlled by dilution and solution during tidal processes. Locally, in the Loire estuary, during seasonal biogeochemical variations, calcite, apatite, and silica may be precipitated (Manickam et al. 1985).

6.1.1.3 Accumulation and Destruction of Sediments Are Important in Deltas Which Extend from the Coastline into the Open Sea

Deltaic accumulation. Deposition in river mouths takes place when sedimentary particles brought down by fluvial channels arrive in a basin with comparatively stationary water masses forming a saline wedge along the bottom (hypopycnal flow, Fig. 6.4A). Sedimentation is intense at the mouth of the river due to the abrupt decrease in the transporting capacity of the waters and the resulting deposition of coarser material carried close to the bottom.

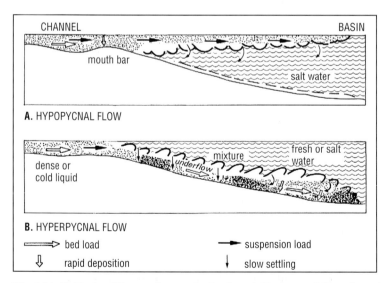

Fig. 6.4 A, B. Types of flow at the mouth of a river. **A** Hypopycnal flow of continental waters dispersing laterally over sea-water forming a salt wedge, with massive sand deposition at the distributary mouth bars; **B** hyperpycnal flow by continental waters specifically heavier than that of the respective basin, leading to outward dispersal along the bottom (after Bates, in Galloway and Hobday 1983)

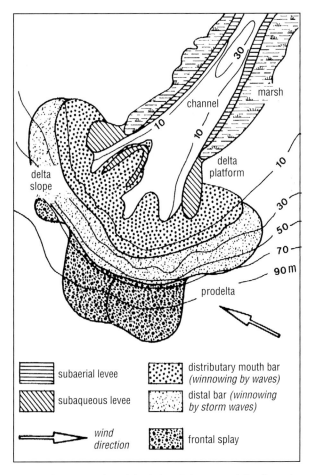

Fig. 6.5. Depositional environments and processes in a delta (after Coleman and Gagliano, in Galloway and Hobday 1983). Wave action leads to a remobilization of sediment parallel to the wind direction

The finer-grained particles transported in suspension pass to the open sea, forming an argillaceous sedimentary fan. In this way, a river mouth bar and a sandy deltaic slope are built up, later to be reworked by normal waves and by storm waves (Fig. 6.5). Farther out, a sandy-silty prodelta is established. When fluvial waters carrying particulate matter flow into a lake (cf. Sect. 5.3), or when very cold or turbid waters run into the sea as in high-latitude deltas, underflows will occur, leading to a rather homogeneous sedimentary fan extending to the open sea. This type of flow is referred to as hyperpycnal or plane jet flow (Fig. 6.4B). Here, there is no river mouth bar.

Overflow of fluvial waters due to collapse or rupture of channel banks during floods is a common process on alluvial fans (cf. Sect. 5.4) and on delta plains. These phenomena contribute prominently to the buildup of sediments

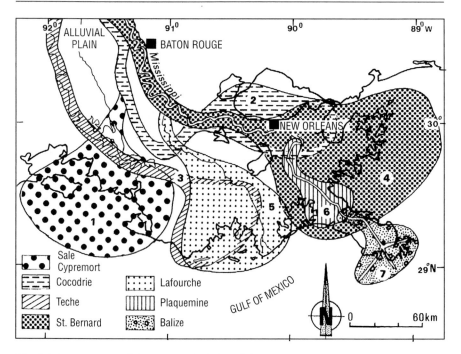

Fig. 6.6. Arrangement of the seven main lobes of the Holocene delta plain of the Mississippi (after Coleman, in Reineck and Singh 1980). The *numbers 1–7* indicate the chronological sequence of the lobes

in deltas, and spillage of the turbid waters leads to formation of second-order clayey-sandy delta complexes (crevasse splays), superimposed over the primary set-up. The Mississippi delta is made up of six vast overflow sheets, formed at different stages of its history which are responsible for the bird-foot type appearance of the active delta lobe (Coleman and Gagliano, in Galloway and Hobday 1983).

Formation of lobes. The successive abandonment of delta channels controlled by the progradation of the sedimentary lobes and by the corresponding loss of transporting capacity of the waters flowing into the open sea leads to the development of new channels and new laterally spreading lobes, which in turn are responsible for further progradation. This seaward progradation through deposition in the river mouth and over the prodelta is assisted by pronounced lateral migration of the main channels and their associated lobes. During the last 6000 years 16 lobes have formed in the Mississippi delta, each of which covered about one-tenth of the total delta area extending laterally over about 240 km. Seven of these lobes have exerted a particularly pronounced influence on the buildup of the delta (Fig. 6.6). The youngest, a bird-foot delta (cf. Fig. 6.1) referred to as No. 7, is the recent, active, Balize lobe.

Destruction of deltas. Waves running at different angles against the coast, remobilize the river mouth deposits and transfer them laterally along the coastline (cf. Sect. 6.2, Fig. 6.5). When reworking is weak, the sand banks resulting from the deltaic progradation are spread out and enlarged. Where wave action is intense, as in front of the Rhone delta, the delta front develops more or less sinuous coastal sand bars roughly parallel to the mean coastline and at right angles to the deltaic channels. Due to the continuous back and forth movement in the channels, tidal currents lead to morphologies opposite to those formed by waves. Channels tend to widen in the downstream direction, banks of silty-sandy material are arranged at right angles to the coastline and the overall structure of the delta approaches that of an estuary like the Fly River of New Guinea (Fig. 6.1). Continuously active marine currents, however, lead to the redistribution of sediments to the front of the prodelta and beyond, sometimes over considerable distances. From the mouth of the Amazon, the North-equatorial current of the Atlantic carries suspended clays northwest, close to the Gulf of Mexico (Gibbs 1974), a distance of more than 2000 km.

Compaction and gravity transport. River mouth deposits and the submerged parts of coastal deltas are particularly vulnerable to gravitational processes. The sands of the river mouth and the delta slope prograde over water-saturated, poorly compacted prodelta clays. These have been deposited rapidly together with organic material subjected to bacterial decomposition, leading to the liberation of gas. Furthermore, the slopes are comparatively steep, especially at the top of the prodelta, and the continental discharge results in an excess of the sediment load in the range of several tens of millions of tons per year. This explains the abundance and diversity of structures resulting from deformation, compaction, and resedimentation, observed in the submarine environment in front of deltas. These processes contribute to the reservoir potential of such sedimentary complexes (Fig. 6.7):

1. rise of plastic masses or mudlumps in mud diapirs through lateral and vertical flow under the weight of sand bars which sink and accumulate locally in great thickness (more than 100 m in front of the Mississippi delta);

2. development of curved faults in the proximal zone of the prodelta, with the concave side pointing in the downslope direction, resulting especially from sedimentary overloading in front of the mouth of active delta channels and leading to mudflow accumulations;

3. subsidence within the prodelta, leading to striated slump faults and to submarine grabens; the extension necessary for this type of deformation appears to be controlled by sudden deep-seated seaward flow ahead of the depocenters or zones of maximum deposition;

4. progressive deformation in open-marine zones of accumulation, with growth faults forming parallel to the coast, and with slumps of great extent which are frequently recorded on deep-sea fans (cf. Sect. 7.1); these curved, or listric faults, tend to become horizontal at depth, displacement along them being compensated on surface by accumulation of sediment.

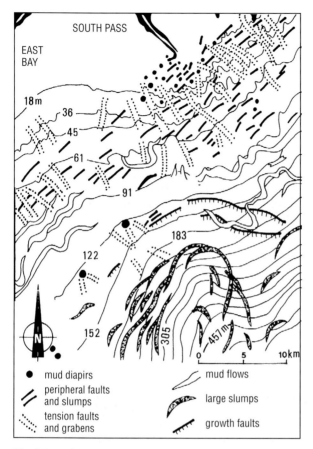

Fig. 6.7. Main types of synsedimentary deformation off the South Pass of the Mississippi delta (after Roberts, Cratsley and Whelan, in Elliott; Reading 1986)

Abandonment of lobes. Deltaic lobes abandoned by lateral migration of the distributory channels are frequently exposed to the action of waves and tides. When they no longer receive continental detritus, they tend to become eroded. This leads to a cyclicity of formation and destruction of sediments through time, accentuated by slumping and subsidence of the abandoned channels, the remains of which eventually may be covered by transgressive marine deposits.

6.1.2 Formation of Delta Complexes

6.1.2.1 Deltas Dominated by Fluvial Process

The best-known example is the recent Mississippi delta (cf. in Blanc 1982; Galloway and Hobday 1983). This delta (Fig. 6.1) consists from the coast seaward of the following successive sedimentary units:

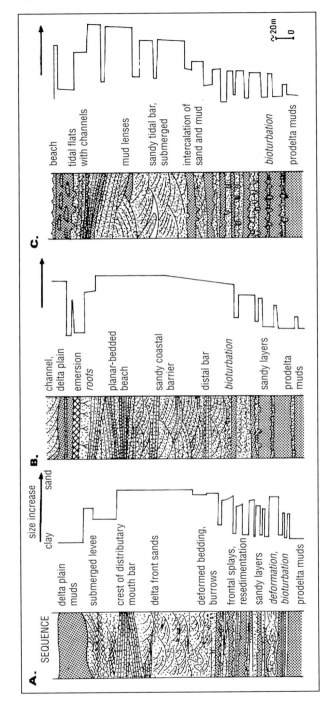

Fig. 6.8 A–C. Examples of deltaic sedimentary sequences. **A** Distributary mouth bar and river-dominated delta; **B** coastal barriers and wave-dominated delta; **C** tidal ridges and deltaic environments off the tide-dominated Colorado River mouth (after Galloway and Hobday 1983)

1. Delta plain: Innumerable channels of low sinuosity with sandy to gravelly bottom, sometimes with climbing ripples and clay pebbles, filled by finer-grained material; less commonly meandering channels with sandy-silty fill; natural banks or bioturbated silty-clayey levees; lateral spill-over deposits or crevasse splays with decreasing grain size along the flow of the flood waters from the point of bank failure; marshes and isolated freshwater lakes separated by delta lobes, and filled by fine-grained sediments rich in organic matter.

2. Distributary mouth bars consisting of well-graded sands with numerous sedimentary structures such as plane bedding, resulting from high-energy flow conditions.

3. Submerged delta front with sandy/silty intercalations, cross-stratification and current laminations, burrows and plant debris, diapiric deformation.

4. Prodelta subjected to wave action, rich in silty-clayey oozes, with few current structures; fauna marine benthic and planktonic, ubiquitous synsedimentary deformation.

5. Clays of the continental shelf in thicker layers, subhorizontal bedding, strong benthic activity (tests, bioturbation), presence of growth faults and slumps, no current structures.

The buildup of a delta dominated by fluvial influences is characterized by seaward *prograding* or off-lapping sedimentary *sequences,* which lead to an increase of grain size from one sequence to the next, a situation refered to as "coarsening-up". It is particularly well developed in sequences at the level of the mouth bars where well-graded sands overlie the silty sands of the delta front and the muds of the prodelta (Fig. 6.8 A). These sequences are frequently terminated by channel or levee deposits, and then by delta plain fills (*agradation*). The abandonment of the deltaic lobes by their feeder channels,

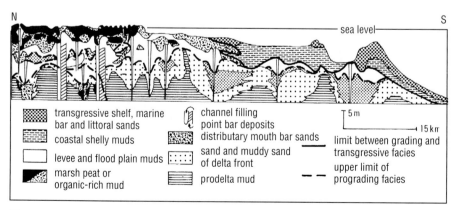

Fig. 6.9. Structure and main sedimentary facies of the Holocene Mississippi delta (constructed from 54 boreholes, after Frazier, in Galloway and Hobday 1983)

compaction, subsidence, and eventually erosion by waves and tides, can lead to marine facies transgressing over the truncated deltaic sequences. In general, the facies resulting from destruction followed by marine *transgression* are larger and exhibit greater lithological homogeneity than the facies of deltaic progradation or agradation (Fig. 6.9).

6.1.2.2 Wave-Dominated Deltas

The Rhone delta is characterized by a fluvio-marine complex subjected to the dominant action of waves (cf. in Oomkens 1970; Blanc 1982). It extends from the Gulf of Fos to Montpellier in irregular lobes shaped by fluvial discharge, marine erosion, and littoral transport, and continues under water in a deep-sea fan down to the center of the Balearic abyssal plain. Its characteristic features are the presence of sandy units emplaced either by the sea as coastal barriers or by the river in the form of channel fills resulting in the maximum development of sand towards the center of the emerged delta (Fig. 6.10). Coastal barriers that progress seaward with the advancing delta are frequently arranged parallel to the coast in contrast to the sand-filled channels. Associated with these facies are silty-clayey deposits of the channel banks, the organic muds of interchannel swamps and lakes with freshwater or brackish faunal elements, the marine or deltaic sediment of contrasting grain size deposited in littoral depressions between the barriers with or without small marginal dunes, and lastly argillaceous marine prodelta oozes largely dispersed by waves, storms and westerly coastal drift. Sedimentary structures are

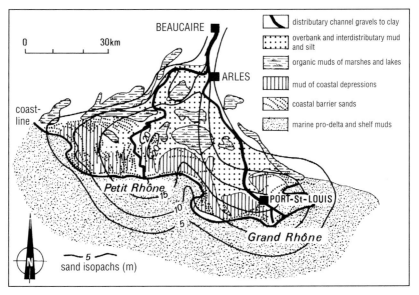

Fig. 6.10. Main depositional environments of the Holocene wave-dominated Rhone delta (after Fisher et al., in Galloway and Hobday 1983)

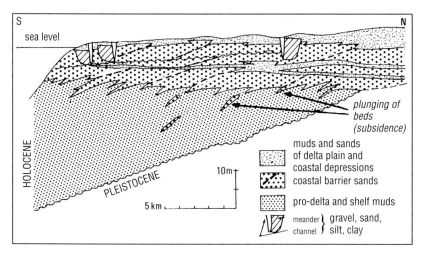

Fig. 6.11. Reconstruction of the Holocene Rhone delta (from data of 9 boreholes, after Oomkens 1970; Galloway and Hobday 1983)

particularly abundant in the sand bodies: wave ripples, planar cross-bedding of coastal barriers, inclined concave cross-bedding of distributary mouth bars with clay pebbles, slump bodies, and decreasing degree of sorting. Hydroelectric power stations of the Rhone basin lead to a retention of sediment far upstream of the delta. It thus does not advance any more, but rather recedes because of coastal erosion (cf. Blanc 1982).

The general structure of wave-dominated deltas is marked by coarsening-upward progradation. This becomes particularly evident in littoral barrier sands: first muds of the shelf and prodelta, then an irregular alternation of sands, silts, and bioturbated muds of the distant delta front, overlain by sand-bars with sedimentary structures resulting from waves and currents, sometimes displaced down-slope en masse by gravity processes, and finally well-sorted plane-bedded beach sands (Fig. 6.8B). These prograding deposits may be obliterated by agradation with heterogeneous deposits resulting from channels and/or the delta plain, with emersion structures and with accumulations of organic material. Subsidence and sedimentary loading in all deltas are indicated by a seaward-increasing plunge of the clayey and sandy beds, by a thickening of the sediment fill (Fig. 6.11) and, locally, by growth faults.

6.1.2.3 Tide-Dominated Deltas

Tidal deltas like those of the Colorado of North America (Meckel, in Broussard 1975; Fig. 6.12A) are similar to true estuaries. Tidal currents arrange the sediment bodies at high angles or even at right angles to the coastline, all the more pronounced when the rate of sediment discharge is low, the tidal differences high, and the slopes regular. The resulting deposits exhibit two sets of characteristic structures:

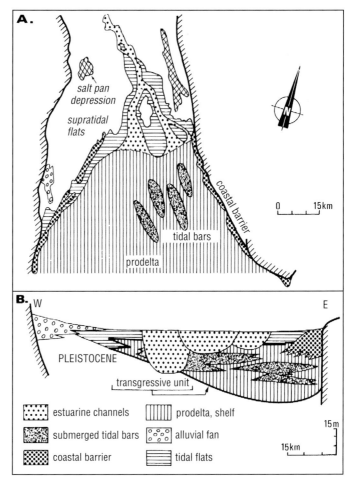

Fig. 6.12 A, B. Main sedimentary facies of the tide-dominated Colorado delta. **A** Plan view and recent deposits; **B** section of Holocene delta across the area of the mouth (from data of 5 boreholes, after Meckel, in Galloway and Hobday 1983)

1. Estuarine channels, separated from each other by intertidal bars of sandy-clayey deposits, exhibit a monotonous sequence of thin, well-sorted, upward-fining sandy layers with current ripples and flaser bedding (sandy ripples with fine-grained clayey intercalations), clay pebbles, etc. These sequences are controlled by the lateral migration of channels and tidal cycles, and are usually covered by more reducing beds resulting from progradation.

2. Tidal ridges in the lower reaches are shaped by the tidal currents from the river mouth deposits. At this position, an upward-coarsening sequence is developed over the bioturbated muds and sands of the prodelta, with numerous current structures and fine-grained lenticular or bedded intercala-

tions (Fig. 6.8 C). Tidal ridges arranged parallel to the direction of tidal currents also form at the mouth of estuaries of French rivers such as the Gironde, the Loire, and the Seine.

During progradation and agradation, the tidal ridges may become covered by intertidal muds, beach sands, estuarine muds and sands, and even organic or evaporitic sediments deposited in coastal marshes or lagoons (Figs. 6.8 and 6.11 B). Synsedimentary deformation appears to be of limited importance in coastal deposits of this type of estuarine deltas where accumulation and lateral growth of the sedimentary bodies frequently remain rather moderate. In contrast, in deltas subjected to a combination of tidal and fluvial processes, subsidence may be pronounced and their economic potential significant. This is the case in the Mahakam delta on the Indonesian island of Kalimantan (Allen et al. 1979) where plant growth on the delta plain represents potential future coal deposits, and where sands of the estuarine sand-banks and tidal ridges have a high reservoir potential for oil.

6.1.3 Fossil Estuaries and Deltas

Fossil estuaries have rarely been described in literature, despite the abundance of recent estuarine complexes. Two factors may be responsible for this:

    1. Recent estuaries result from the Holocene eustatic transgression of the sea into the lower reaches of fluvial valleys and thus constitute structures of short duration on the geological time scale. They are prone to erosion and downstream fluvial plugging when the terrigenous discharge increases or when the sea-level recedes, forming a sequence of descending (offlap) wedges. During marine transgression or when the terrigenous discharge is diminished, they experience marine plugging. This sequence of ascending wedges is called onlap.

    2. Estuarine deposits consist of various proportions of marine and fluvial compounds, and are frequently difficult to characterize. One has to look for indicators of tidal currents and fluvial floods, for mineralogical markers of these diverse environments, and for fossils indicative of gradients in salinity from freshwater to open marine conditions. The thickness of estuarine series is usually small as is their lateral extent, thereby limiting opportunity for identification.

Fossil deltas have been described from geological formations of various ages where they are of considerable vertical and lateral extent, being locally of high economic interest (cf. Elliott, in Reading 1986). The recognition of fossil delta complexes relies on the nature and arrangement of the sand bodies (channels, distributary mouth bars, coastal barriers, tidal ridges, etc.; Table 6.1), and on the relative development of sedimentary sequences such as progradation, agradation, abandonment, and even marine transgression. The characteriza-

Table 6.1. Main characteristics of delta deposits (after Galloway and Hobday 1983)

	Dominant influences		
	River	Waves	Tides
Lobe geometry	Elongated or lobate	Sinuous or arcuate	Estuarine to irregular
Bulk composition of sediments	Clayey or heterogeneous	Sandy	Clayey to sandy
Characteristic facies	Distributary mouth bar and delta front sheet sands	Coastal barrier sands	Tidal bar sands
Channel type	Low sinuosity, suspended-load to mixed load	Sinuous, bed-load to mixed-load	Variable, depending on tide energy

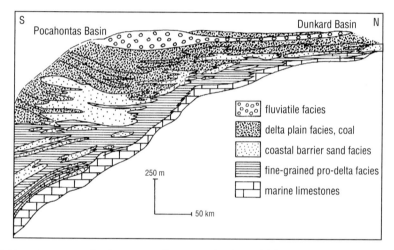

Fig. 6.13. Sedimentary succession in a prograding wave-dominated delta complex at the edge of the Carboniferous Pocahontas and Dunkard basins, West Virginia and Pennsylvania/USA (after Ferm and Cavaroc, in Leeder 1982)

tion of the above two groups of factors, together with the localization of zones of synsedimentary deformation (slumps, various types of faults), usually allows the identification as to whether river, waves, or tides dominated the formation of a delta, and the reconstruction of the structure of the ancient delta complexes. Thus, the Carboniferous Pocahontas and Dunkard basins of Virginia and Pennsylvania/USA show the development of a wave-dominated delta with coastal barriers above neritic marine limestones (Ferm 1974; Fig. 6.13).

Numerous deltaic series exhibit a succession of superimposed sedimentary cycles. Each cycle corresponds to a prograding lobe with the following sequence: clayey prodelta, sandy delta front, deltaic plain with different

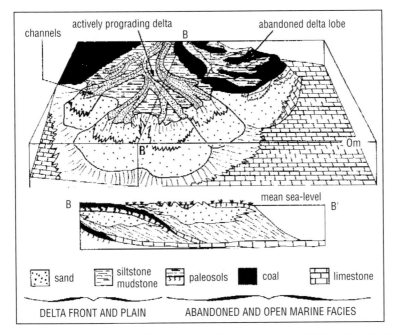

Fig. 6.14. Reconstruction of formation and abandonment of delta lobes, North American Carboniferous (after Ferm, in Elliot; Reading 1986)

lithofacies, sandy fluvial alluvial plain. Each lobe advances seaward above the preceding lobe. Numerous deltaic sequences are covered by marine transgressive deposits. Each sequence naturally is somewhat irregular, being subjected to lateral migration, erosion, and to abandonment of lobes. Prograding and agrading facies of certain North American Carboniferous deltas, for instance were successively abandoned and eroded under strong fluvial influence, and then covered by continental sediment, or superseded by weathering profiles (Ferm 1974; Fig. 6.14). On abandonment of a lobe, the cessation of the detrital discharge favors the seaward formation of marine limestones.

6.2 Littoral Environments

6.2.1 Hydrosedimentary Mechanisms

6.2.1.1 Introduction

The littoral domain extends from the coastal plains, dunes, and cliffs to the sea bed at a depth of several tens of meters. Marked by pronounced local and seasonal variations of morphological, hydrological, hydrochemical, and biological parameters, this environment, more than any, is controlled by the

Littoral Environments   185

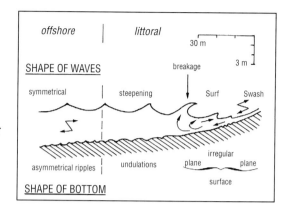

Fig. 6.15. Shape and movement of waves against the coast, together with ripple morphology on exposed beaches of Oregon/USA (after Clifton, Hunter and Philipps, in Leeder 1982)

action of waves and tides. *Waves* caused by wind, as well as the residual swell, lead to simple oscillating currents in the open sea without defined transport direction of the water particles. In contrast to this, the movement of waves and swell results in a landward displacement of the water: the velocity of the waves and wavelength decreases due to friction along the sea bed, whereas curvature and height increase (Fig. 6.15). As it approaches the coast, the crest of a wave becomes progressively more asymmetrical until the frictional forces lead to breakage. Several breakages can follow each other to the beach if the slope of the bottom is low. After breaking, the water progresses against the bottom in complex waves (surf zone) which run over the beach in a back-and-forth movement (swash/backswash), the period of which is the same as that of the original breaking waves.

Tides are deformations of water masses under the influence of the gravitational pull exerted by the moon on its course around the earth, and by the sun on moon and earth on their joint passage around the center of our planetary system. Tidal currents over the continental shelves suffer complex deformations and amplifications controlled by irregularities of the coastlines and submarine morphology. The amplitude of the tides on average increases towards the borders of the large oceanic basins with their topographic narrows and slightly inclined bottoms (Fig. 6.16). Depending on their amplitudes during the equinoxes, the tides are subdivided into micro-tidal (below 2 m) as along the Mediterranean coasts, mesotidal (2–4 m), and macro-tidal (above 4 m) as along the French coast of the Atlantic and the English Channel. Within the same region, the tides may range from micro- to macro-tidal. This is the case for the North Sea coast of the Netherlands, Germany, and Denmark where macro-tidal zones in the estuaries of the Elbe and Weser gradually pass to micro-tidal zones along the interrupted barrier islands to the north and west (Fig. 6.17). Detailed information on the movement of marine waters and on the associated sedimentary mechanisms are presented by Komar (1976, 1983), Davis (1978), Walker (1978), Guilcher (1979), Leatherman (1979), Reineck and Singh (1980), Blanc (1982), Galloway and Hobday (1983), and Elliott (in Reading 1986).

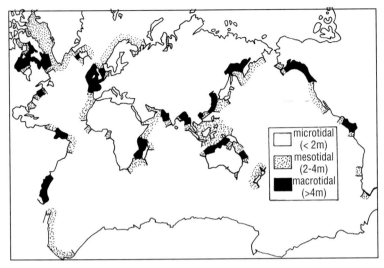

Fig. 6.16. Amplitude of tides around the margins of the world's oceans (after Davies 1973)

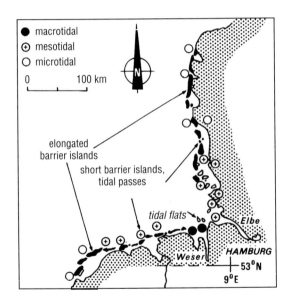

Fig. 6.17. Relation between tidal range and coastal morphology in northwestern Europe (after Hayes, in Reading 1986)

6.2.1.2 Coastal Waves

Close to low-lying, frequently sandy coastlines, the crests of waves and swells approach each other and are deformed by friction along the sea bed. The crest lines of the waves tend to parallel the position of the depth contours, as the advance of the water is slowed down more over shallow bottoms than over deeper ones. Coastal refraction of waves leads to various modifications, depending on submarine relief, wave length and frequency, and orientation

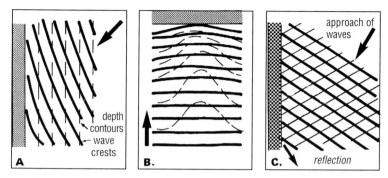

Fig. 6.18 A–C. Refraction of waves on shallow coastline (**A**+**B**) and reflection from steeper coastline (**C**) (after Leeder 1982). **A** Waves at an angle to coastline and depth contours; **B** waves at an angle to depth contours; **C** waves at an angle to the coastline

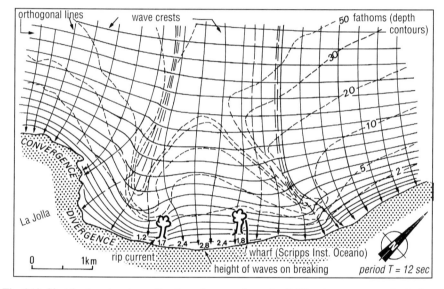

Fig. 6.19. Sketch plan showing refraction of waves along the Californian coast near La Jolla. Deformation of the wave crests is controlled mainly by convergence along the La Jolla promontory and by divergence above two smaller submarine canyons (after Munk and Traylor 1974; Shepard and Inman, in Komar 1983)

against the coastline (Fig. 6.18 A and B). In general, waves advancing against the coast acquire a convex curvature in bays and over deeper waters. This leads to a loss of energy due to the divergence of the orthogonal crest lines, and to the deposition of fine-grained sediment. In contrast to this, the waves acquire a concave shape against capes and over shallow bottoms, the crest lines tend to converge, and the resulting concentration of forces leads to active erosion and deposition of coarser-grained material. The advance of waves and surf against the coast results in a number of phenomena of erosion, transport, and sedimentation, important for coastal management and the respective protective measures. They are often of such complexity that it is necessary to prepare graphic or photographic maps illustrating the situation (Fig. 6.19).

After breaking against low-lying sandy coasts, waves control the position of coastal currents, the nature and sedimentary potential of which depend mainly on the angle between wave crests and the coastline:

1. When the waves are breaking parallel to the coast and with a uniform amplitude, the water retreats without lateral displacement until it is taken up by the subsequent wave in a swash-backswash motion. More frequently, the amplitude of the breaking waves varies along the coast (cf. Fig. 6.19), and in zones of higher breaking waves the water runs off laterally in a symmetric fashion, forming an independent cell. The water retreats seaward over the bottom along axes oriented mainly at right angles to the coast, taking with it the sediment eroded from the beach (Fig. 6.20 A). The intensity of the resulting *rip currents* depends on the topography of bays and the meterological regime. These phenomena control the formation of beach cusps, leading to transport of large quantities of littoral sand to the sea.

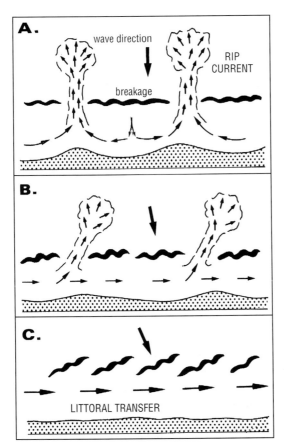

Fig. 6.20 A–C. Types of littoral currents. **A** Breakers parallel to the coast, lateral migration and formation of rip currents at right angles to it, leading to littoral erosion; **B** intermediate situation; **C** breakers at an angle with the coast, responsible for long-shore drift and erosion

2. When the waves are breaking at an angle of above 5°–10° to the coastline, reflux tends to take place parallel to the coast and the alternation of flux and reflux leads to continuous longitudinal migration of the coastal waters (Fig. 6.20 C). This results in a locally important *littoral transfer* of sediment which is responsible for the formation of numerous sediment bodies, from littoral spits and shore-connected banks to barriers closing off bays or estuaries (Davies 1973).

3. When the angle between plunging waves is below 5°–10°, a combination of rip currents and littoral transfer occurs (Fig. 6.20 B).

Along more sloped, generally rocky coastlines, the greater depths of the bottom are not conducive to refraction or breaking of the advancing waves, the crests of which then remain parallel and almost rectilinear until they encounter the beach. This results in a *reflexion* of the waves (Fig. 6.18 C) which also can cause notable littoral transfer. The energy of the waves which here are little retarded by bottom friction, frequently is considerable, especially when storms combine with exceptional tides. Erosion is accelerated particularly where the bottom gradient is rather steep close to the coast. The retreat of the Atlantic cliffs of Normandy and Picardy may reach 2 m/year. They constitute a good example of erosive phenomena which advance through mechanical abrasion and collapse along fracture planes, just as well as through biological activity. The progress of erosion is inhibited by the masses of fallen rocks, the more resistant components of which are transformed by abrasion to the pebbles of littoral coarse clastic deposits. The sediment-binding action of organisms such as algae and polychetes worms also protects beaches against erosion.

6.2.2 Detrital Environments

6.2.2.1 Main Types of Sedimentary Environments

Detrital sediments deposited essentially along low-lying coastlines and in bays, are mostly of a silico-clastic nature, less frequently of carbonates. They characterize three main types of environments:

1. *Beaches* represent littoral zones exposed directly to the dominant action of waves, accompanied by tides of different amplitude. As the energy of the waves increases in the direction of the coast, so does the mean grain size of sedimentary material deposited by them. This is especially pronounced along the straight sandy beaches of France (English Channel near Calais, east coast of Bay of Biscay, Gulf of Lion, etc.). From the coast seaward, the typical profile of a low-lying sandy beach (Fig. 6.21) consists of:

a) the terrestrial zone, frequently made up of aeolian dunes;

b) the back-shore, inundated by the sea during storms and exceptionally high tides, corresponding to a supratidal environment with its characteristic supralittoral biological communities, or biocoenoses (Pérès 1961);

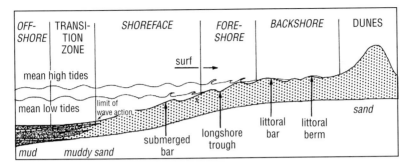

Fig. 6.21. Typical morphological units in section of a sandy beach

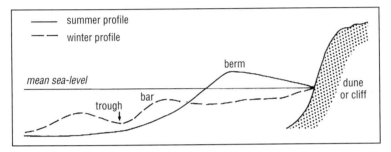

Fig. 6.22. Characteristic winter and summer profiles of a beach (after Komar 1976)

c) the beach proper, or fore-shore situated between the mean high- and low-water levels, the intertidal environment with medio-littoral biocoenoses;

d) the shore face, the upper portion of which falls dry during exceptionally low tides, whereas the lower portion is subjected to the influence of waves along the bottom (the start of infratidal environment with infralittoral biocoenoses);

e) a transition zone subjected to the influence of storm waves, passing seaward into the argillaceous sediments of the continental shelf proper (offshore environment, circalittoral biocoenoses).

A beach profile frequently exhibits pronounced seasonal variations: Oscillating summer waves lead to the accumulation of a convex littoral berm, whereas the breaking winter waves carve out an erosive concave profile resulting in a sequence of submerged bars and troughs parallel to the coast (Fig. 6.22).

2. The *tidal flats* represent vast, shallow and little-inclined areas subjected predominantly to tidal action along macro- and mesotidal coastlines. The energy of the waves and the mean grain size of the bottom sediment decrease in the direction of the beach due to the progressive slackening of the currents and the inability of the waves to break. The subhorizontal bottoms, usually rich in fine-grained deposits, are dissected by large, meandering channels. Ex-

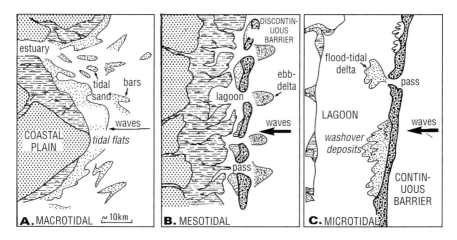

Fig. 6.23 A–C. Morphological control exerted by tides on littoral deposits, accompanied by moderate wave action (after Hayes, in Leatherman 1979). **A** Macrotidal regime with minor wave action; **B** mesotidal regime; **C** microtidal regime

amples are the littoral zones developed outside the estuaries of northwestern Europe (Elbe, Rhine, Schelde).

3. *Bars and littoral barrier islands* result from the regular accumulation of sediment along coastlines without pronounced tides, such as Wadden Sea of the northern Netherlands or the coastline of southwestern Denmark (Fig. 6.17) and also certain section of the French coast (Verger 1983). In micro-tidal environments they constitute linear rarely interrupted bedforms susceptible to being washed over by storm waves. Behind them, littoral lagoons are formed. In mesotidal environments, they are less elongated and frequently breached by tidal inlets through which the water of the rising and falling tides will pass. Behind these barriers, marshes dissected by channels are developed. In the micro-tidal environment, tidal deposits form mainly during rising water and behind the littoral bars in flood-tide deltas. In the mesotidal environment, deposition takes place mainly during falling tide and on the seaward side of the bars in ebb-tide deltas. The main morphological characteristics of beaches subjected to tidal action are compiled in Fig. 6.23.

6.2.2.2 Modern Littoral Environments

On *sandy beaches exposed to waves,* a landward increase in hydrodynamic regime and grain size is accompanied by mainly planar sedimentary structures as the littoral bars exposed to higher flow regimes are approached. Strata resulting from successive sedimentary accumulations exhibit different degrees of inclination, depending on the morphology of the bars and the run-off direction of the falling tides (Fig. 6.24). They are sometimes interrupted by cross-bedding, smaller in the troughs between the bars and larger along the tidal

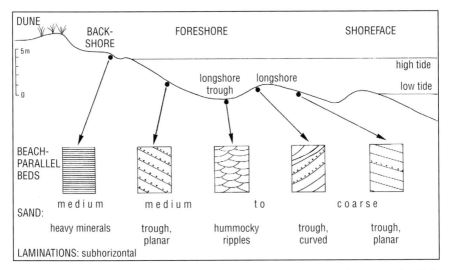

Fig. 6.24. Sedimentary structures on the la Coubre beach, north of the mouth of the Gironde/SW France (after Allen et al. 1981)

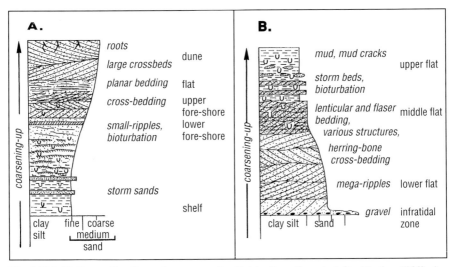

Fig. 6.25 A, B. Prograding littoral sequences in siliciclastic sediments (after Tucker 1982). **A** Wave-dominated beach or littoral barrier; **B** tide-dominated tidal flat

channels. In the direction of the beach, structures indicative of emersion (hummocky cross-beds, escape structures) and eventually the large cross-stratifications of dune complexes are encountered.

Seaward, the sands become increasingly finer-grained, grading into argillaceous sands, and eventually into bioturbated muds, unless intense tidal currents preclude the deposition of fine-grained material, as in the English Channel. When sedimentation dominates over erosion, the littoral sedimentary complex starts to prograde seaward, leading at a given point to an upward-coarsening sequence (Fig. 6.25 A).

Sediment bodies resulting from littoral transfer and connected to the coast at one of their extremities, form a number of different types: Sand spits and associated curvated refraction banks extending from a cape and fed by fluvial material; shore-connected banks and tidal ridges controlling the lateral migration of tidal channels, accompanied by symmetric erosion on wind-exposed coastlines like Picardy (Despeyroux 1985); tombolos or littoral bars joining an island to the continent like those of Quiberon (South Brittany) and Giens (Provence, Southeastern France); strings of sand bodies closing off bays and leading to the formation of natural polders (cf. Davis 1978). A particularly spectacular example is presented by the sand bar displacing the mouth of the Senegal River to the south, and by successive flint pebble bars migrating northward along the Atlantic coast of the French province of Picardy from the chalk cliffs of Ault to the bay of the Somme (Fig. 6.26).

Emerged sand bars form mostly as a result of sedimentary growth of submerged bars. Thus, during progradation they exhibit a sequence similar to

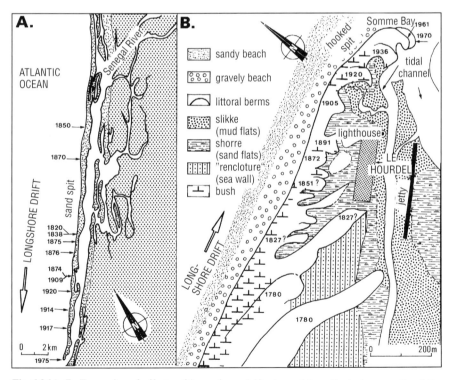

Fig. 6.26 A, B. Examples of effects of long-shore drift: **A** sand bar displacing mouth of the Senegal (after Zenkovich, in Friedman and Sanders 1978); **B** successive pebble bars south of the Somme bay, Picardy/France (modified after Dallery 1955)

that of a classical beach. One also finds the fine-grained facies of the backshore lagoon (upward-fining), or sandy facies resulting from washover by storm waves, or erosive facies overlying a flood-tide or ebb-tide delta platform (cf. Figs. 6.23 B and 6.30). Behind the back-shore lagoons, in tropical climates, mangrove swamps can develop, as well as surfaces with algal mats or evaporite deposits (cf. Sect. 6.2.3.2). The littoral barrier islands exhibit quite different evolutionary paths with time, depending on whether they are subjected to transgression, regression, or simple agradation:

1. In a transgressive environment, sandy marine deposits overlie with sharp contact the fine-grained, organic-rich lagoonal deposits behind the littoral barrier washed over during storms. Marine erosion and inland migration of the emerged barrier lead to the destruction of the upstream deposits as on Sapelo Island of Georgia/USA (Fig. 6.27 A). The inland advance of erosion

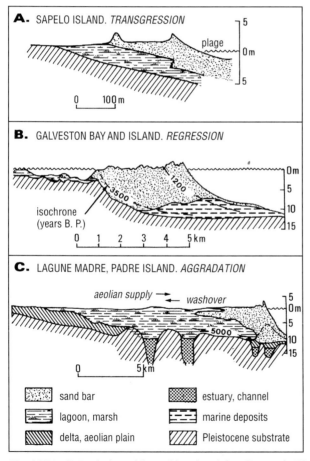

Fig. 6.27 A–C. Evolution of littoral barriers (after Curray, in Blatt et al. 1980). **A** Transgression (destruction); **B** regression (progradation); **C** agradation (plugging)

sometimes results in outcrops of ancient clayey lagoonal deposits or peats on the higher shore. This is the case north of the Gironde estuary in the Coubre Bay of southwestern France (Allen 1981) and on numerous beaches near Boulogne on the English Channel. These deposits, transgressing in the form of sedimentary wedges, present a so-called "on-lap" facies.

2. In a regressive environment, progradation results in the accumulation of a negative, or coarsening-upward sequence, advancing seaward, a structure called "off-lap". The resulting deposits grade from infratidal muds to intertidal sands and then supratidal shelly sands as on Galveston Island/Texas (Fig. 6.27 B).

3. Agradational regimes correspond to the presence of stable coastal barriers, behind which the lagoons are gradually filled by washover deposits resulting from storms (tempestites), by expanding flood-tide deltas, or by aeolian material blown in from inland, e.g., Padre Island, Texas (Fig. 6.27 C).

Tidal flats are vast surfaces little influenced by tidal currents, over which the landward decrease in energy leads to increasingly finer-grained sediments. An

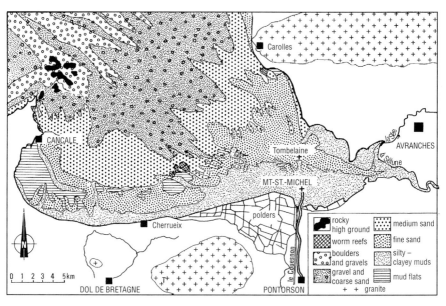

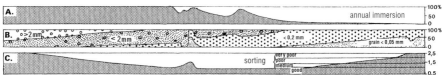

Fig. 6.28 A–C. Schematic distribution of recent sediments in Mt. St. Michel bay, northwestern France with variations of certain parameters along a WNW-ESE section (after Larsonneur 1975; Caline 1981). **A** Rate of annual submersion; **B** grain-size distribution; **C** sorting (Inman standard deviation)

example in temperate climates is the bay of Mont-St. Michel of northwestern France. It is exposed to intensive filling as a result of the Holocene rise in sea-level and of human activities, such as dykes, canals, polders, flood locks, mussel farming, etc. The deposits successively encountered in the direction of the coast (Fig. 6.28; Larsonneur 1975; Caline 1981) are: boulders, gravel, partly biogenic or gravelly coarse sands, quartzose and bioclastic sands of medium to fine grain size with numerous sedimentary structures, litho-bioclastic and bioturbated fine-grained sands and silts, silts and clayey silts of the emerged mud flats (slikke), sand flats (shorre), and exposed salt flats.

Northwest of the bay, large sand beaches exposed to waves are bordered by littoral dunes. Meandering and braided tidal channels dissect this immense flat. Towards the south-central part of the bay a large reef built by sessile tubiform worms (*Sabellaria alveolata*) is developed, behind which fine-grained sediments are deposited. Mud (= "tangue") of the upper mud flats and the sand flats of the bay, is a detrital, poorly graded (sandy-silty-clayey) but predominantly fine-grained gray material with a mean size between 0.9 mm and a few micrometers. It is rich in bioclastic carbonate fragments (40%–50%) and rather depleted in clay minerals. The thin bedding results from deposition by successive tides, exhibiting a gradation from the lower, coarser and lighter-colored base, through normal size grading into an upper, fine-grained and darker lamina. The progradation of tidal flat deposits leads to normal upward-fining sedimentary sequences when passing from the lower to the upper intertidal environment (Fig. 6.25 B).

Coarse or shelly sand bars, or *cheniers,* are locally observed on flat silty-clayey beaches, especially around estuarine bays. They represent elongated ridges up to 50 km long at a mean height of 3 m, resting with a sharp contact on a clayey base. These unique bedforms which are particularly well developed along the Louisiana coast of the Gulf of Mexico (Fig. 6.29), result from storm deposits in zones of usually fine-grained sedimentation where clayey material is supplied by littoral transfer or tidal currents. Belts of shelly material similar to cheniers are encountered in the eastern portion of the bay of Mont-St. Michel (Caline 1981) where they can be deposited by storm waves up to the mud flats.

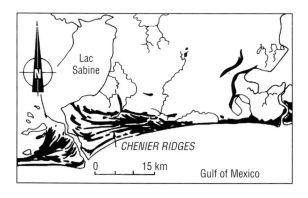

Fig. 6.29. V-shaped arrangement of cheniers deposited by storm waves on silty-clayey tidal flats of Louisiana, with material derived from the Mississippi by lateral transport (after Gould and McFarlan 1959)

6.2.2.3 Ancient Littoral Deposits

The identification of coastlines in ancient sedimentary sequences is of special importance for the reconstruction of the paleogeographic evolution of an area, for the recognition of tidal regimes, of the influence of wave action compared to that of tides, and of the paleo-depth and the associated environments. Such basic investigations are also of practical use. Buried littoral sand bodies frequently represent good hydrocarbon reservoirs due to their high primary porosity and good permeability. The silty, organic-rich lagoonal sediments constitute potential source rocks, whereas the intercalated crevasse splays and flood-tide deltas can serve as traps. Littoral barriers may enclose exploitable epigenetic concentrations of uranium like the Tertiary of central Texas and can also lead to the formation of peats and coal in the coastal swamps extending on their landward side. Last, but not least, marine beaches locally harbor important placer deposits of, e.g., diamonds (southwestern Namibia), or of heavy minerals such as platinoids, monacite, rutile, ilmenite, tourmaline, cassiterite, etc. In literature, a large number of reconstructions have been presented (cf. especially Reineck and Singh 1980; Miall 1984; Elliott, in Reading 1986) which are based on the interpretation of the succession of various lithofacies and sedimentary structures of mechanical or biological origin (Fig. 6.30).

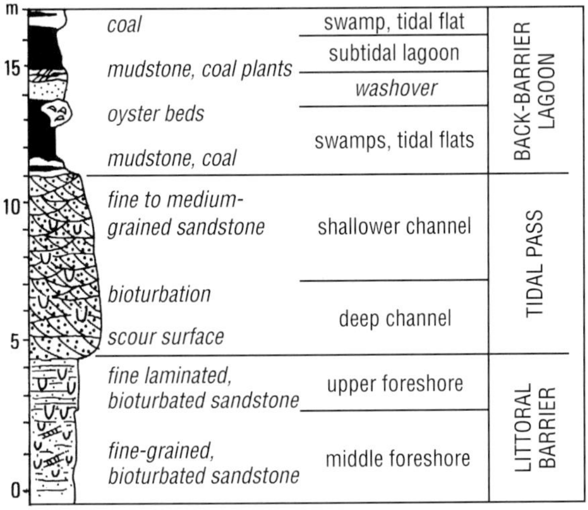

Fig. 6.30. Stratigraphic sequence in Canadian upper Cretaceous Blood Reserve Formation (St. Mary/Alberta) attributed to a prograding complex of littoral barrier – tidal pass – coastal lagoon in a mesotidal environment (after Young and Reinson, in Leeder 1982)

Lithology	Facies	Environment
coal	swamp, tidal flat	BACK-BARRIER LAGOON
mudstone, coal plants	subtidal lagoon	
	washover	
oyster beds		
mudstone, coal	swamps, tidal flats	
fine to medium-grained sandstone	shallower channel	TIDAL PASS
bioturbation		
scour surface	deep channel	
fine laminated, bioturbated sandstone	upper foreshore	LITTORAL BARRIER
fine-grained, bioturbated sandstone	middle foreshore	

6.2.3 Carbonate and Evaporite Environments

6.2.3.1 Arid Tidal Flats and Sabkhas

The coastal deposits of arid tropical regions have been studied in particular along the southern part of the *Persian Gulf*, the coast of the Arab Emirates (Trucial Coast), and in Shark Bay, Western Australia (cf. Purser 1983). In the

198   Marine Coastal Environments

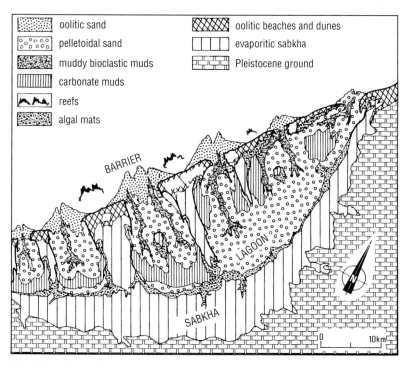

Fig. 6.31. Distribution of main recent types of sediment along the southern coast of the Persian Gulf in Abu Dhabi (after Purser and Evans, in Purser 1983)

Persian Gulf, the excess of evaporation compared to precipitation (1500 mm/a against 40–60 mm/a) leads to an increase in the salinity of the interstitial waters of the inter- and supratidal zones. Thus, intertidal sedimentation along the coast of Abu Dhabi is dominated by stromatolitic algal mats up to 2 km wide, dissected by a network of tidal channels and basins (Fig. 6.31). The seaward adjoining infratidal lagoon is protected from wave action by a discontinuous chain of reefs and by a littoral sand barrier and oolitic dunes. Oolitic sands form mainly in deltaic fans extending seaward between the various coastal islands of Abu Dhabi (Loreau and Purser 1973; Fig. 6.32). These are zones of strong agitation where tidal currents debouching into the sea encounter coastal currents and littoral transport. Other places of oolite formation are tidal barriers, certain portions of beaches, the back-barrier lagoon and sheltered tidal flats (Purser 1983). The lagoonal deposits which consist essentially of pelletoid sands, carbonate muds, and bioclastic argillaceous sands deposited in channels, are periodically reworked by storm waves, and shed over the stromatolitic zones where their material is trapped by the algal cover. In the landward direction, in the supratidal zone up to 16-km wide sabkhas are encountered, in which gypsum, anhydrite, and Mg-rich carbonates are precipitated. Progradation leads to the burial of the intertidal algal mats by the evaporites which are rapidly altered by compaction, bac-

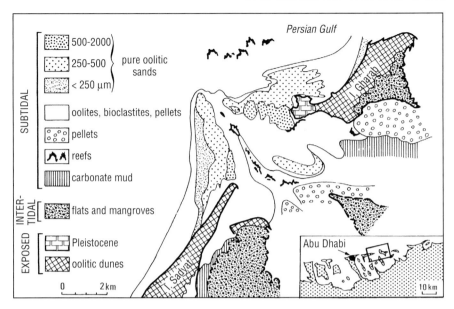

Fig. 6.32. Sedimentary facies in an oolitic delta of Abu Dhabi between the islands of Sadiyat and Gharab (after Loreau and Purser, in Purser 1983)

terial activity, and the growth of interstitial gypsum (Park 1977). The prograding sequences exhibit a typical regressive situation, from infratidal oozes and aragonitic sands over degraded algal mats to supratidal evaporites (Fig. 6.33 A). There is some morphological convergence between the coastal carbonate barrier systems of the Abu Dhabi-type and siliciclastic barriers of the Wadden-Sea type of the northern Netherlands: coastal barriers with lagoons, tidal channels, tide deltas, sedimentary back-bar ripples, and sedimentary structures of the inter- and supratidal zones. The morphological convergence underlies the importance of mechanical processes during sedimentation in the two cases described.

In *Shark Bay*, which is also situated in an arid environment with 230 mm/a precipitation and 2200 mm/a evaporation, two types of algal growth can be observed:

1. Continuous algal mats, trapping all sorts of carbonate grains in zones of low energy; fine laminations with fenestral structures predominate in submerged lagoons, whereas poorly laminated, wavy algal mats form above the mean high-tide level.

2. Columnar lithified stromatolites form reefs in submerged areas of elevated energy; the resulting structures are associated with littoral barriers rich in cross-bedded mollusc debris.

The supratidal zone is occupied by evaporitic deposits, locally covered by intraclast sands during storms. The infratidal zone and the shelf carbonate as-

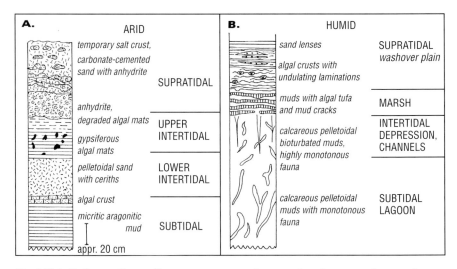

Fig. 6.33 A, B. Prograding sedimentary sequences in coastal carbonate and evaporite environments. **A** Tidal flats and sabkhas in arid climates, Abu Dhabi/Persian Gulf (after Till, in Reading 1986); **B** tidal flats in humid environment, west coast of Andros Island/Bahamas (after Hardy and Ginsburg, in Leeder 1982)

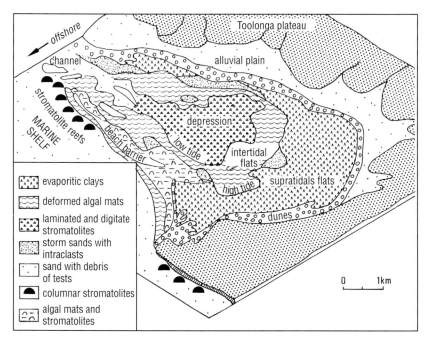

Fig. 6.34. Sedimentary associations in the Hutchinson Gulf of Shark Bay/Western Australia (after Hagan and Logan, in Leeder 1982)

sociation (cf. Sect. 6.3) are connected with the coastal complexes through channels filled by carbonate sand consisting of skeletal debris (Fig. 6.34).

6.2.3.2 Tidal Flats and Lagoons in Humid Regions

Littoral carbonate deposits of tropical humid regions, characterized by the absence of evaporites, are well represented by the western side of the Bahamian island of *Andros* (Hardie 1977) or by the Florida lagoons (Ginsburg 1964). The elevated and variable precipitation (640–2000 mm/a) leads to considerable fluctuations in salinity between 5‰–42‰, which are responsible for the faunistic monotony over the tidal flats. The littoral zone on Andros is marked by a low tidal range (0.5 m) and frequent storms. Its sediments occur in three associations which tend to prograde seaward (Fig. 6.33 B):

1. Infratidal pelletal argillaceous carbonate sands which are strongly bioturbated, especially by the crustacean Calianassa, and separated from the open sea by coarser-grained facies such as littoral bars, tidal channels, and storm-generated washover fans.

2. Fine-grained pelletoidal carbonate oozes of the intertidal surfaces, rich in burrows, mollusc shells, and desiccation structures. The tidal flats are dissected by a dense network of channels (1–100 m long, 0.2–3 m deep) exhibiting little lateral migration in contrast to the channels in temperate regions (cf. Sect. 6.2.2.2). The bottom of these channels is covered by coarse sediment, with mangroves and hemispherical stromatolite bodies occupying the accompanying levees.

3. Algal mats and encrustations of supratidal and terrestrial marshes. The stromatolites are frequently lithified to algal tufas by high-Mg calcite. Drier periods result in numerous desiccation structures, whereas during storms sand tends to become incorporated in irregular, 1–10-cm thick laminations within the algal mats.

The Florida Gulf which is sheltered against the open sea by a chain of Pleistocene carbonate platforms or cays, is covered by carbonate oozes with few sandy grains, exhibiting structures of the wackestone or mudstone type. These particularly fine-grained carbonate sediments of the infratidal zone frequently show spongy oncoids made up by filamentous bluegreen algae. The little diversified fauna abounds in small burrowing bivalves, while marine plants and algae are also frequent. A unique aspect of this region is represented by carbonate mud banks which at a height of about 2 m may be several kilometers long. They are locally outlined by emerging islets, arranged in a network (Ginsburg 1964). The formation of these mud banks appears to be connected to the action of storm waves on unconsolidated intertidal material and to the trapping of calcareous oozes by sea-grass (Thalassia).

6.2.3.3 Ancient Littoral Deposits

Ancient calcareous or evaporitic littoral formations have been described mostly from arid environments, where they generally pass seaward to formations deposited on carbonate platforms (cf. Sect. 6.3). A number of examples were compiled by Miall (1984) and Till (in Reading 1986). The main characteristics of the carbonate deposits (Leeder 1982) as compared to those of the siliciclastic environments, are the following:

1. common in intertropical regions;
2. size of the particulate compounds reflects more the type of biological or biochemical origin than the hydrodynamic conditions;
3. clayey-silty oozes result mainly from the decomposition of calcareous algae;
4. modification of sedimentary environment by organogenic structures, without the necessity for changes of hydraulic regime;
5. important and rapid modifications of particulate compounds and matrix, widespread cementation during diagenetic evolution normally starting shortly after deposition.

The lower Jurassic of southern England, and the lower Purbeckian in particular, represent a littoral environment in an arid region comparable to the

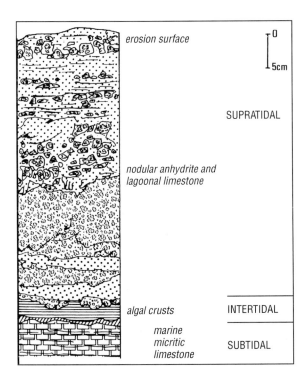

Fig. 6.35. Sabkha sequence prograding over littoral carbonates of lower Purbeckian age in England (Warlingham borehole/Sussex) after Shearman, in Till, Reading 1986)

modern coastal sabkhas of the Persian Gulf-type (cf. Fig. 6.33) and their carbonate margins (West 1965; Shearman, in Reading 1986). Drill cores from Sussex, south of London, show complete repetitive sequences in which evaporites prograde over lagoonal limestones and algal mats (Fig. 6.35). Farther to the west, in Dorset, four successive sedimentary facies have been identified:

1. limestones consisting of pellets and intraclasts with foraminifera and stromatolites deposited in a moderately hypersaline infra- to intertidal environment;

2. stromatolitic and oolitic limestones with gypsum pseudomorphs after calcite, attributed to a strongly hypersaline intertidal environment;

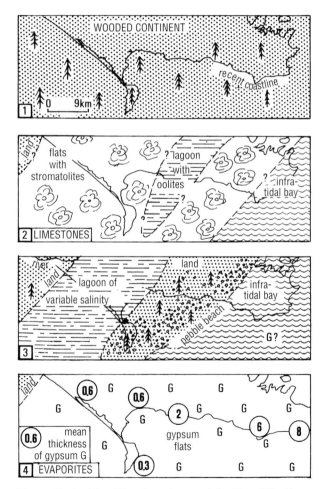

Fig. 6.36. Paleogeography of lower Purbeckian littoral environment in Dorset/Southern England (successive stages *1–4* after West 1965; Till, in Reading 1986)

3. more or less brecciated limestones replacing abundant primary anhydrites precipitated in a hypersaline inter- to supratidal environment;

4. ostracod limestones without gypsum pseudomorphs, as evidence for the re-establishment of more normal, marine inter- to infratidal conditions.

The recognition of these different facies within successive beds, and their correlation between the various outcrops and boreholes, allow deciphering the evolution of the coastline during the late Jurassic and reconstruction of the paleogeographic development of a littoral environment which in places was carbonatic, in others evaporitic, and even emerged (West 1976; Fig. 6.36).

6.3 Shelf Environments

6.3.1 General Hydrosedimentary Mechanisms

More than any other sedimentary environments, the continental shelves reflect the *complexity of the exchange processes* characterizing the external geodynamic realm. The particles transported by creeks, rivers, or wind have to pass through a number of sediment traps such as estuaries, deltas, lagoons, and tidal flats, until eventually they arrive on the shelf. Here they are subjected to complex hydrodynamic actions exerted by currents resulting from waves, tides, and density differences, before a fraction of them reaches the continental slope and the oceanic basins. The position of this zone of sedimentary transit and exchange is shifted by numerous phases of reworking during the course of geologic history. This is a consequence of fluctuations in sea-level which, in turn, tend to complicate the distribution of sedimentary bodies on the shelf. Certainly this applies to glacio-eustatic fluctuations during the uppermost Cenozoic. Until about 11 000 years ago, the sea-level was situated at the outer edge of the shelves. The Flandrian transgression advanced over irregular coastal surfaces underlain by alluvial, glacial, and aeolian deposits. Intrusion of the sea into the lower reaches of alluvial valleys led to their upstream plugging by sediments, whereas the numerous relic deposits outcropping on the shelves were subjected to reworking by hydrodynamic agents during the Holocene. However, the fairly recent age of the shelves, and the particular conditions of sedimentation on them limit their applicability for the interpretation of ancient shelf deposits. Detailed data have been presented by authors such as Swift (in Burk and Drake 1974), Vanney (1977), Boillot (1979), Reineck and Singh (1980), Blanc (1982), Galloway and Hobday (1983), Purser (1983), Reinson (in Walker 1984), Johnson (in Reading 1986), and Sellwodd (in Reading 1986).

The shelf environment covers the area between the littoral zone below 10 m depth, i.e., the zone of influence of normal wave action, and the top of the continental slope between 100–250 m depth. Where there is no clear break in the slope along the continental margin, the lower limit of the shelf is arbitrarily set at a depth of 200 m. The break in slope itself can be determined by isostatic warping and faulting, active sedimentary progradation, or by trapping of sediments behind tectonic horsts as along the East Pacific, behind limestone reefs in the Red Sea, or behind salt diapirs in the Gulf of Mexico. The width of the shelf varies between 1 km and about 1500 km, depending on the topography of the continental margin and the tectonic context. Passive-margin shelves of the Atlantic type are generally much wider than those of active margins of the Pacific type. In their majority, modern continental shelves are only moderately subsiding and their sediment cover is much less than that of some ancient shelves. Siliciclastic deposits of the Scottish Precambrian attain a thickness of 5000 m, those of the Cambro-Ordovician of South Africa some 2000 m, whereas the shelf carbonates of the lower Cretaceous of southern Europe reach several hundreds of meters in thickness. The sandy deposits frequently encountered in these environments of transition, constitute excellent hydrocarbon reservoirs where covered by argillaceous sediments or forming sedimentary wedges.

The dynamic processes affecting the shelves are quite diverse in nature, being dominated by waves and tides in the inner zone, and by density and oceanic currents in the outer zone (Fig. 6.37). *Tidal currents* develop with increasing intensity around restricted zones with small tidal movements, called amphidromic points. These points, around which the flood isochrons or cotidal lines are rotating during a tidal cycle, are located especially in the centers of marine basins. In shallow areas like the North Sea several amphidromic points may be developed (Fig. 6.38). The tidal waves are usually asymmetric, the flood component being faster than the ebb component, leading to the preferred transport of sediment particles against the beach. *Storm-generated waves* can also appreciably affect the shelf deposits. Ordinary, wind-generated waves here are without effect, except occasionally over the crest of megaripples and submerged sand ridges. During storms, the influence of waves can extend down to a depth of 200 m, as for example in the Bay of Biscaye (Castaing 1981). In the northeast Pacific, on the shelf off the state of

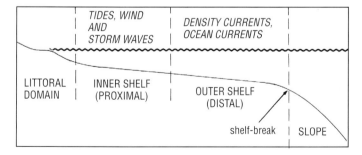

Fig. 6.37. Zonation of continental shelf and dominant hydrodynamic processes (after Galloway and Hobday 1983)

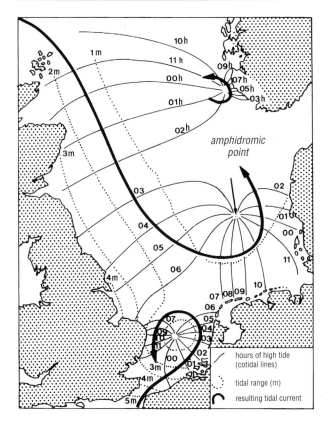

Fig. 6.38. General tidal regimes of the North Sea (after Harvey, in Johnson, Reading 1986)

Washington in the northwestern United States, transport of sediment by waves has been noted over 5 days per year at 167 m depth and over 53 days at 75 m depth (Sternberg and Larsen 1976). The effect of storm waves is frequently very important in the respective deposits or tempestites, and takes the form of long symmetrical and bifurcating ripples. In contrast to the effect of the tides, these waves lead to active erosion of parts of beaches and thus to a growth of the shelf. Liquefaction of surface layers of sediments and reworking of terrigenous coastal muds may also be initiated by these waves, leading to the seaward deposition of mass flows or of laminites (cf. Chap. 7). The effects of these various processes become enhanced when they are superimposed on each other, such as when a storm or hurricane occurs during exceptionally high tides, and when approaching the littoral environment.

Density currents on shelves result from the surficial supply of freshwater by rivers which are responsible for the deposition of terrigenous argillaceous fans, as well as from intense evaporation in hot climates which supports the precipitation of carbonates and early diagenetic reactions. Stratification ac-

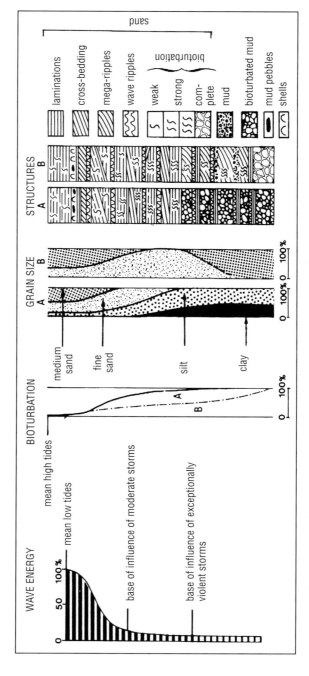

Fig. 6.39. Sedimentary zonation and benthic activity on continental shelves marked by absence (*A*) or presence (*B*) of strong tidal currents (after Reineck and Singh 1980)

cording to densities, referred to as pycnocline, leads to the establishment of a thermocline as in lakes (Sect. 5.3.1.2). This may be accompanied by the appearance of a halocline, i.e., a discontinuity in salinity. Strong landward winds cause forceful wave currents, that is, a rush of water mobilized against the beach and backflow or undertow above the bottom which is responsible for coastal erosion and the accumulation of sand and gravel farther out to sea. The effect of these undertows is enhanced when combined with storms. Seaward winds push the surficial water layers out to sea, a movement compensated for by the rise of abyssal waters frequently enriched in nutrients. This process is referred to as upwelling. Such waters support the profusive development of plankton, the formation of oxygen-deficient marine beds, and the accumulation of organic remains, siliceous tests, phosphates, etc. in the resulting sediments. Semi-permanent oceanic currents affect the most distal portions of certain continental shelves where they lead to transport of sedimentary material parallel to the coastline. The Agulhas current in the western part of the Indian Ocean for example passes over the outer portion of the shelf off Southern Africa, transporting fine-grained sand southward (Flemming 1980). This longitudinal transport frequently terminates at submarine canyons, from the top of which particle flows become directed seaward (e.g., off the Californian coast; Moore 1969).

Except for reef structures, *biological bottom activity* is inversely proportional to the intensity of the hydrodynamic processes involved. Bioturbation is intense in clayey distal shelf deposits sheltered from strong tidal currents like the terrigenous coastal muds of the Gulf of Lion in the northwestern Mediterranean and along the north coast of the Gulf of Mexico. In contrast to this, bioturbation is weak in sandy and gravelly sediments on shelves subjected to intense tidal currents or longitudinal oceanic currents. Good examples of this occur in the eastern part of the English Channel, the southern North Sea, and off the coast of southwestern Africa (Fig. 6.39). When the shelves are repeatedly covered by tempestites or by periodites, the benthic organisms buried start to recolonize the new sediments from the burrows in the upward direction, in contrast to normal sediments which are colonized from the top down. The distribution of the invertebrates is controlled in particular by water depth and the availability of oxygen. Oxidized clayey sediments are colonized by numerous bivalves, echinoderms, and polychaete worms which are filtering or grazing clayey material. In oxygen-depleted environments, mobile limivores with few hard constituents (polychaete worms) prevail, whereas in reducing environments only few suspensivores are encountered. Marine vegetation represents effective sediment traps like the posidonia meadows of the Mediterranean. It has to be reiterated, that on the distal parts of shelves with weak sedimentation, in certain micro-environments, glauconite and berthierine are frequently formed, especially from fecal pellets and in the interior of tests.

6.3.2 Detrital Environments

6.3.2.1 Sedimentary Facies

In sandy facies developed especially on shelves subjected to active tidal currents (Fig. 6.40), a number of different types of bedforms are encountered. *Sand ribbons* are bedforms of only several decimeters thickness which locally extend to 20 km length and 0.2 km width. They are sometimes identical to each other in shape and spacing. They form at depths of 20–100 m under the influence of strong tidal currents with velocities of the order of 1 m/s which probably contain a turbulent component. Although they are mainly encountered on gravelly bottoms with only a thin sandy cover, they may also develop at the extremities of sand-banks subjected to strong tidal currents as in the eastern English Channel. *Sand waves* are dune-shaped bedforms 3–15 m high, with a wavelength of up to 600 m over sandy shelf areas with strong tidal currents below the wave base of the region. The shape of these sand waves is

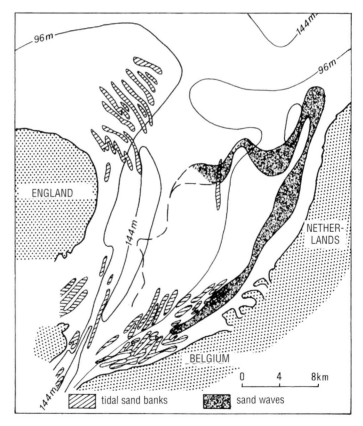

Fig. 6.40. Main tidal ridges and sand waves in the southern North Sea (after Houbolt, McCave, in Galloway and Hobday 1983)

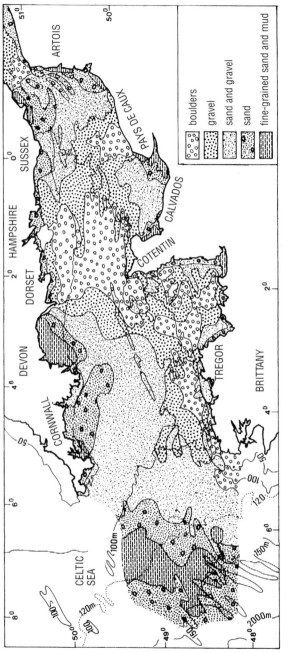

Fig. 6.41. Schematic distribution of sediments in the English Channel (after Larsonneur et al. 1982)

symmetrical or asymmetrical, depending on whether the vector resulting from the alternating currents is zero or not. The internal structure of these bedforms is still rather poorly known. They appear to be dominated by large cross-beds advancing by avalanching in the down-flow direction, with smaller stratification oriented in the opposite direction superimposed on the upstream flank (Galloway and Hobday 1983). *Tidal ridges* are bars or banks elongated parallel to the direction of the residual tidal currents. They are rather abundant in the southern part of the North Sea where they consist of medium-grained, well-sorted sands. They occur in groups up to 40 m high, 2 km wide and 60 km long, spaced at intervals of 5–12 km. Their transverse profile is asymmetrical, with an internal structure made up of stratifications slightly inclined in the direction of the steeper side which itself shows inclinations of less than 6°. The tidal ridges appear to be mostly in equilibrium with the present marine situation and frequently appear to resemble ancient littoral bars successively shaped by tidal currents which established themselves during the post-glacial Flandrian transgression (cf. Swift, in Burk and Drake 1974). An example of tide-dominated, coarse-grained clastic shelf sedimentation is provided by the English Channel (Fig. 6.41), the sediments of which are dominated by boulders, gravel, and sand (Larsonneur et al. 1982). The deposits are mainly bioclastic in the west-central part between Brittany and Cornwall, becoming more siliciclastic in the east and west. The distribution of the sediments is controlled by the intensity of the tidal currents which is at a maximum between the two amphidromal points of the region. Thus coarse pebbles accumulate on the right-hand side of the Cotentin peninsula and the Straits of Dover where the currents reach velocities of up to 3 knots (about 1.5 m/s) and locally even 8 knots, such as off Cape La-Hague. In contrast to this, the sediments become finer-grained away from these zones of high energy, grading into muds in sheltered bays like Devon, Mont St. Michel, the Seine estuary, or below depths of 100 m in the direction of the Atlantic in the Celtic Sea.

Muddy facies are developed off coasts with weak tidal processes as a result of the decrease of the wave energy towards the outer shelf (Fig. 6.39). This is the case on the shelf of the Bering Sea, the southwestern Gulf of Mexico, the Bay of Biscaye, and the Gulf of Lion, where heavily bioturbated terrigenous coastal muds form transition to open-marine hemipelagic deposits. Despite the seaward increase of the fine-grained facies, hydrodynamic processes are expressed in a number of different ways. Storm waves can rework the fine-grained sediments and supply coarse littoral sands to the outermost edge of the shelf (Sect. 6.3.1.2). At the same time, the break in slope between shelf and continental slope is frequently marked by an increase in current activities which favor the development of filtering animals such as crinoids, and, in contrast to the deposition of recent sediments, permit the uncovering of relic sands of upper Pleistocene age as in the Gulf of Lion (Aloisi et al. 1975). Furthermore, processes of hydrodynamic sorting are identified in the same way as in areas of weak currents and winnowing, i.e., by variations in the relative abundance of clay minerals. In general, the content of smectites and other ex-

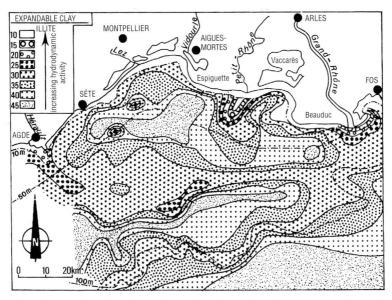

Fig. 6.42. Relative abundance of swelling clay minerals (smectite and associated mixed-layers) in surface deposits of the eastern Gulf of Lion (after Arnoux and Chamley 1974)

pandable minerals, the deposition of which is favored by calm conditions, increases seaward from the coast by *differential settling,* at the expense of illite and kaolinite (Gibbs 1974, 1977). On the western shelf portion of the Gulf of Lion, smectites and the associated random mixed-layer minerals permit the distinction of two different zones (Fig. 6.42):

1. A zone rich in expandable minerals, favorable for decantation: This takes place through differential sedimentation in front of the Greater and Lesser Rhone; beyond the break in gradient between shelf and continental slope; in strips parallel to the coast over the inner shelf; and over the outer shelf where notable bottom currents are absent.

2. A zone low in expandable minerals where hydrodynamic processes are more pronounced: This takes place close to river mouths where fluvial currents are dying off; at the break in gradient between shelf and continental slope; along currents at right angles to the coast; over the mid-portion of the shelf subjected to winnowing action by the westward "liguro-provencal" current which is strengthened by Rhone-induced currents; in a circular coastal strip underlying an eastward countercurrent; and in local areas with turbulent currents.

6.3.2.2 Sedimentary Sequences

The succession of different types of sediments at a given point on the continental shelf is mainly controlled by variations in thickness of the overlying

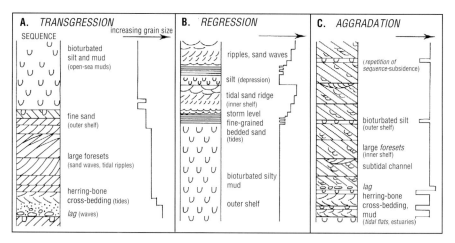

Fig. 6.43 A–C. Schematic sedimentary shelf sequences. **A** Transgression in a tide-dominated region; **B** regression, with progradation over bottoms influenced by tidal currents and sometimes storms; **C** aggradation through cyclic plugging resulting from repeated sudden subsidence in a mixed tide- and storm-influenced region (after Galloway and Hobday 1983); *lag* large particles left in place; *foreset* group of cross-beds forming on the lee side of current ripples

water column with time. These in turn are determined by pulsations in the expanse of the oceans by glacio-eustatic processes, by the importance of subsidence and its fluctuations, and by the tectonic situation along the respective ocean margins. The successions thus deposited may be transgressive, regressive, or aggradational in character (Fig. 6.43) with a number of modifications, depending on whether the shelf is subjected mainly to tidal processes, storm waves, or plain decantation.

Transgression sequences are characterized by a general upward decrease in mean grain size (Fig. 6.43 A) as a result of the rise in sea-level and a decrease in the solid load supplied by rivers. The successive sedimentary layers, which progress landward in the form of an "onlap", start off frequently as gravel and sand deposited by waves, then grade into herringbone cross-bedded sands (tides) and large sandy bedforms (sand waves, tidal ridges), and eventually fine-grained sands with scattered sedimentary structures, passing into intensely bioturbated terrigenous coastal muds. The presence of upward-directed burrows and hummocky cross-bedding allows the distinction of storm deposits from normal sediments. The Holocene transgressive sequences represent widespread examples of such sedimentary successions. In ancient sediments, transgressive sequences up to 200 m thick, favored by subsidence along active margins, are observed in the Cretaceous of the Circumpacific belt (Bourgeois 1980).

Regression sequences are characterized by a seaward sedimentary progradation ("offlap") with a general upward increase in grain size (Fig. 6.43 B). The locally glauconitic, bioturbated muds of the outer shelf grade upwards into sandy beds of the inner shelf, in places interrupted by tempestites. These are overlain by cross-bedded sands of successively submerged ridges, alternating with deposits of depressions (bioturbated silts) and of channels (concave ripples). The resulting deposits are not very thick and thus readily fossilized, except for areas of strong subsidence. The majority of well-known ancient shelf sequences are from such a prograding system. An excellent example is found in the epicontinental Cretaceous of North America, especially in Wyoming and western Canada (cf. Brenner 1978). Investigations in the state of compaction of such sediments during deposition permit the reconstruction of the rate of subsidence and palaeodepths.

Aggradation sequences result from sedimentation by repeated blockages on a shelf subjected to irregular abrupt subsidence. They comprise, above a littoral facies of tidal flats which preceded the subsidence, a sequence of bioturbated muds and silts of the outer shelf, alternating in a more or less regular pattern with cross-bedded sands and channels of the inner shelf (Fig. 6.43 C). Depth indicators do not exceed the upper and lower limits of the shelf, suggesting a delicate eustatic equilibrium. The Jura quartzites in the upper Precambrian of Scotland (Dalradian) constitute spectacular fossil examples of an aggrading sequence in which sands have been deposited by storm waves and tides over a subsiding shelf with a thickness of about 5000 m (Anderton 1976).

6.3.2.3 Paleogeographic Reconstructions

Investigation of the distribution of lithofacies, grain size, sedimentary structures, fossils, and the degree and type of bioturbation in boreholes and sections in a certain area, permits the reconstruction of fossil continental shelves (cf. Johnson, in Reading 1986). The Dalradian of Scotland represents an ex-

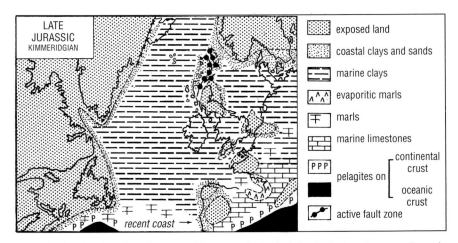

Fig. 6.44. Shelf facies of northwestern Europe and North Atlantic during the upper Jurassic (after Hallam and Sellwood, in Reading 1986)

ample of a shelf dominated by sandy facies (Anderton 1976). The Jurassic of the northwestern Atlantic and parts of the North Sea are dominated by argillaceous facies, which become increasingly more abundant from the Toarcian to the Kimmeridgian when they formed hydrocarbon reservoirs (Hallam and Sellwood 1976; Fig. 6.44). The argillaceous oozes represent an open-shelf facies with numerous trace fossils, reduced facies improverished in burrowing animals, and a laminated bituminous kerogen-rich facies containing rare, specialized faunas. The sandy facies result from the littoral shelf margins, whereas to the south marls and then open-marine limestones make their appearance.

6.3.3 Carbonate Environments

6.3.3.1 Introduction

There are three main types of carbonate environments, as a function of the hydrodynamic context (Wilson 1975):

1. Shelves exposed to hydrodynamic actions facilitate the formation of indurated masses of carbonate muds and of various types of debris trapped by organisms, forming discrete structures on the continental slope right up to the photic zone (Fig. 6.45 A). These muds mounds are also common in fossil se-

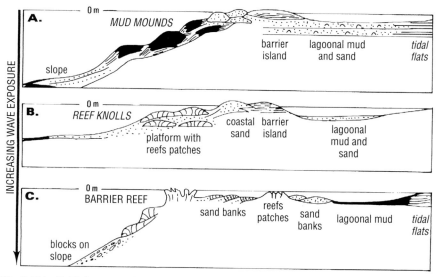

Fig. 6.45 A–C. Main types of carbonate complexes on the shelf (after Wilson 1975). **A** Mud mounds on sheltered shelf; **B** isolated reef knolls on slightly sloped shelf with moderate hydrodynamic activity; **C** reefs on exposed shelf

quences, their constituent organisms comprising sponges and algae (Cambro-Ordovician), bryozoans and algae (Ordovician to Permian, including the Waulsortian facies of the European and North-American Carboniferous), crinoids (Silurian to Carboniferous), rudist bivalves (Cretaceous), and then marine phanerogamian plants (Cenozoic). There are very few examples of mud mounds in recent environments. However, certain structures accumulating in Shark Bay and Florida (Thalassia seagrass), and on the continental slope off the Bahamas permit interesting comparisons.

2. Shelves exposed to moderate wave action favor the formation of independent structures arranged parallel to the coastline in areas of very low gradients (Fig. 6.45 B). These are subcircular, fairly small reef knolls, consisting of colonies of madreporians, sponges, and columnar stromatopores, separated from each other by sand-sized bioclastic detritus. These structures, which at present are developed especially in the lagoons within ring atolls like those of the Bikini archipelago of the South Pacific, are known in particular from the rudist reef belts of the upper Cretaceous in the Gulf of Mexico (cf. Wilson 1975).

3. Shelves extensively exposed to waves and storms are the sites of formation of reef complexes in the true sense. They form close to the coast as fringing reefs or at some distance from the coast as relatively elongate barrier reefs, or as circular atolls around subsiding volcanic islands (Fig. 6.45 C). Characterized by the abundance of hexacoralline madreporians, coral reefs are widespread in recent tropical environments (e.g., Bahamas, Florida, Australia, Madagascar; cf. Purser 1980, 1983) and in certain fossil sequences (cf. Wilson 1975). The development of the madreporians, the construction role of which is aided by encrustating organisms like algae and bryozoans, and by various producers or binders of sediment, is controlled by the following main prerequisites:

a) shallow water (less than 100 m), warm (18°–36° C) and of normal salinity (27‰–40‰);

b) clarity of water;

c) abundant food supply of zooplanktonic origin;

d) presence of a sufficiently hard and stable substrate permitting settlement by organisms. The classical corals, also referred to as hermatypic, are characterized by a symbiosis with microscopic zooxanthellic algae in the photic environment. They are distinguished from the ahermatypic corals which can form smaller, permanently submerged reefs in deeper waters (below 1000 m) with lower temperatures (down to $-1°$ C).

6.3.3.2 Examples of Modern Reefs

The large Tulear reef is situated on the southwestern coast of Madagascar along the Tropic of Capricorne (Picard 1967; Weydert 1973; Blanc 1982). Facing the Moçambique Channel and predominantly southwesterly winds, it attains a width of 1–5 km at a distance of 3–5 km from an arid coast. Its length

Shelf Environments 217

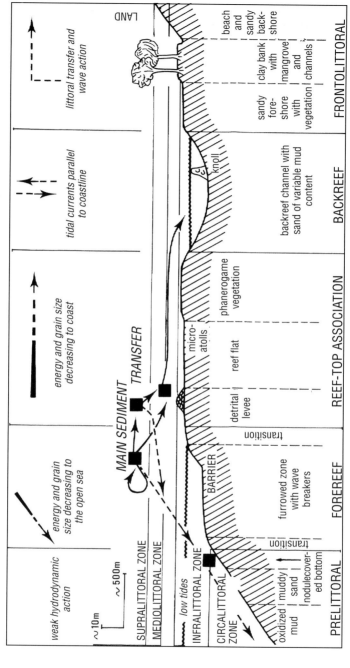

Fig. 6.46. Zonation of Tulear reef and its associated complexes (after Picard 1967; Weydert 1973; Blanc 1982)

of about 18 km is limited by two river mouths, to the north and south respectively. The following zonation from the sea to the coast is observed (Fig. 6.46):

Prelittoral zone (biocoenoses of the circalittoral environment; Picard 1967). This is the upper region of the continental slope and of the non-reef-bearing shelf. It receives little supply of particulate matter from the actual reef itself. It is made up of open-marine oxidized muds (300–90 m depth), detrital muddy sands (90–45 m), and then sandy-clayey substrate with nodules of calcareous red algae (melobesiacea) at depths of 45–25 m. Locally the sandy substrate of the reef complex crops out.

Fore-reef zone. This is the most active and dynamic part of the reef, made up of an alternation of ridges of coralline growth with erosional depressions facing the direction of the dominant waves. It extends from about 25 m depth up to the lowest positions of the sea-level (lower infralittoral environment), constituting the actual reef barrier. It feeds the bioclastic sedimentation mainly in the back-reef area.

Reef-top zone (upper infralittoral environment). This shows a decrease of hydrodynamic activity and thus also of grain size in the direction of the coast. From a region of transition or breaking of waves during high tides, the following zonation is observed:

1. detrital reef crest, formed by the accumulation of blocks broken from the ridge and channel part of the reef during storms and subjected to strong biodegradation;

2. a wide inner reef flat with alternating coralline ridges and tidal channels filled by coarse bioclastic sands (biosparites, intrasparites); living madreporians, scattered or sometimes forming micro-atolls, develop over a substrate of earlier dead corals;

3. an accumulation of fine sand forming a relief resulting from sediment binding by locally concentrated marine phanerogamian plants (especially Thalassia, Cymodocea); this zone is marked by meandering tidal channels which are the loci of accumulation of sands with large foraminifera living in the marine plants (Marginopora, Amphistegina, Sorites...).

Back-reef zone (infralittoral environment). This is the 2–15 m deep back-reef channel washed by tidal currents parallel to the coast. Its floor is covered by more or less argillaceous bioclastic sands. There are also some isolated coral reefs or knolls, together with marine plants. As in the fore-reef zone, little material issues from the actual reef complex itself.

Frontolittoral zone. This is subdivided in the landward direction into three successive zones:

1. a low foreshore with seagrass and clayey sands (upper infralittoral);

2. the littoral mangrove swamps with organic to terrigenous muds bound by algal banks and with channels (Lyngbya, Vancheria; mediolittoral);

3. the shore and backshore with bioclastic and siliciclastic sands (medio- and supralittoral environment) marked by the presence of phreatic springs.

6.3.3.3 Structureless Carbonate Platforms

Carbonate platforms with few continuous reef complexes are developed mostly in warm climates in a number of different morphological environments: platforms in an oceanic environment like the Bahamas, margins of infracontinental marine basins like the western part of Persian Gulf, margins of pericontinental basins like the northwest coast of Australia, and the ones fringing the Yucatan (Mexico) and Florida peninsulas (Purser 1983). Carbonate platforms in temperate climates, however, are rather rare. They exist especially in the Mediterranean Sea between Tunisia and Libya where a number of different sediment belts succeed each other in the seaward direction as follows: littoral quartz sands, calcareous muds with marine plants and algae-covered bottoms, bioclastic sands, muds with increasing clay content, planktonic limestones, calcareous relic sands, and eventually argillaceous carbonate oozes (Burollet and Winnock 1979).

The Bahama Grand Banks, situated some 100 km southeast of Florida, is a wide carbonate platform active since the Cretaceous. Protected from terrigenous supply by the depression of the Florida Straits, this region is characterized by a considerable development of aragonitic oozes behind the Pleistocene reef barriers exposed to the predominantly easterly winds (Fig. 6.47). West of Andros Island, for example, there is a fairly shallow zone (3–5 m) extending over about 120 km. This zone is subjected to tropical temperatures varying seasonally between $21°–29°$ C, a rather high rate of precipitation of 100–150 cm per year, and the agitation of the waters increases towards the western edge of the platform. The tidal currents are moderate (tidal range below 1 m), the salinity increases to 37–46 g/l in the direction of the flats, and the zone shows a preferential development of algae with aragonitic tests (Purser 1983). The main types of sediments encountered from the open sea in the direction of Andros Island are:

1. Deeper carbonate sands and muds, resulting from intense resedimentation (fans at the base of the slope), from size sorting (fine-grained turbidites and laminites along the edge of the basins), and from decantation (pelagites of the slope and the open ocean).

2. Bioclastic sands of the platform edge, rich in debris derived from algae, benthic foraminifera, molluscs, and corals. These sands form a thin layer in the downwind direction and a thicker layer along the eastern coast exposed to the trade winds, being associated here with small, isolated coralline complexes of low relief (patch reefs).

3. Oolitic sands of the platform edge, consisting locally of more than 90% of well-sorted ooids which make up the most important recent example of this type of sediments in the world. These white sands with large cross-beds occur in elongated dunes emergent at low tide along the western edge of the platform, in large areas at the bottom of bays such as Ocean Tongue and Exuma Sound, or in flood-tidal deltas along the eastern boundary of the shelf (Fig. 6.47).

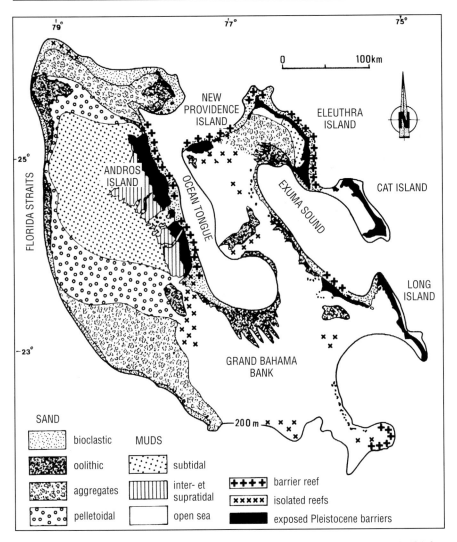

Fig. 6.47. Distribution of the main sediment types on the Grand Bahama Bank shelf (after Purser 1983)

4. Carbonate sands of the inner platform which are intensely bioturbated, but contain few current structures. They are made up of oolites in the shallower exposed zones (e.g., south of Andros Island); coarser sands, consisting of poorly lithified aggregates with little matrix in the less agitated sectors; and micritic pelletoidal sands over sheltered subtidal bottoms.

5. Lagoonal aragonitic muds. These are whitish, and unstratified due to strong bioturbation by molluscs and annelids, with the result that they are rich in soft pellets producing a somewhat lumpy texture. They are widespread in

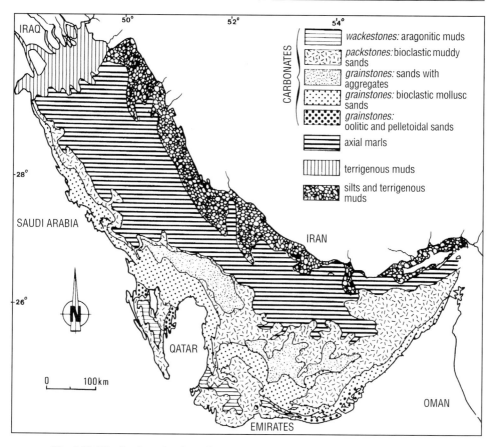

Fig. 6.48. Distribution of main sediment types in the Persian Gulf (after Purser 1983)

shallow infratidal zones in the immediate downwind vicinity of Andros Island. Whitings sometimes observed in the waters overlying these muds appear to be favored by mobile faunal activity, such as fish close to the bottom, rather than to chemical precipitation of carbonates.

6. Muds and sands on tidal flats where fine-grained deposits are interrupted by ingression of storm-mobilized sand and silt, and by meandering tidal channels rich in sand derived from gastropods and lithoclasts. The supratidal marshes contain laminated oozes and sand with stromatolites, desiccation structures, and air-filled cavities, the so-called bird's-eyes.

The Persian Gulf constitutes another recent example of a non-built carbonate platform. It is characterized by water depths generally below 100 m, a salinity locally reaching 40‰, an abundance of carbonate hardgrounds with a specialized epifauna (evidence for weak sediment productivity), and a low rate of deposition. The shallow environments (5–30 m) are dominated by grainstones made up of oolites and pellets, by worn bioclastic debris from molluscs, foraminifera, algae, and sometimes corals, and by aggregates (Fig. 6.48). Deeper down, the hydrodynamic sorting diminishes, organic debris is more angular and embedded in a low-Mg calcitic matrix which becomes more pronounced in the seaward direction (packstones, wackestones). The true carbonate sediments disappear along the central axis of the Persian Gulf at a depth of about 80 m where marls with a strong terrigenous aeolian component take over. The detrital contribution increases again in the direction of the northeastern margin of the gulf along the Zagros Mountains of Iran and off the Shatt-el-Arab delta in the extreme northwest (Purser 1983).

6.3.3.4 Ancient Carbonate Platforms: Morphology and Lithofacies

Like their modern equivalents, fossil reef complexes exhibit zones built up by organisms and zones of sediment accumulation under the influence of currents. A *bioherm* is a lens- or dome-shaped limestone structure built in situ by madreporian corals, algae, bryozoans, sponges, polychaete worms, bivalves, and other organisms. They are generally surrounded by *associated cross-bedded formations* which are discontinuous and varied in nature. They form close to a bioherm as a result of its dismantling and include reef breccias, rudists, and various calcarenites. A *biostrome* consists of well-stratified layers made up of calcareous coralline particles sorted and washed together by currents. It does not show a direct spatial connection to a bioherm. Within a

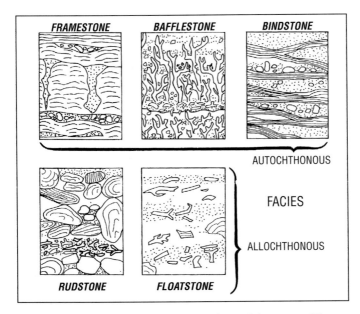

Fig. 6.49. Main facies of allochthonous and autochthonous reef limestones (after Embry and Klovan 1971)

biostrome, the calcarenite layers can alternate with coarser micritic oozes, such as rudist fragments, as a result of variations in hydrodynamic conditions during deposition of the carbonate sediments.

The facies in reef limestones depend on whether the particles have been displaced as in biostromes and the surrounding sediments, or remained in situ as in a bioherm. Embry and Klovan (1971) defined two groups of facies, based on a study of upper Devonian reefs in southern Canada (Banks Island):

1. Limestones consisting of authochthonous components comprising framestones made up of massive in-situ fossils which control the organization of the deposit, and bafflestones in which the skeletons left in place result from upward-growing and sediment-binding organisms. A third type are the bindstones with tabular and lamellar fossils responsible for the retention of sediment and the formation of encrustations.

2. Limestones made up of allochthonous components are rudstones consisting of grains in contact with each other, and floatstones in which grains of above 2 mm size, not in contact with each other, account for more than 10% of the rock (Fig. 6.49). As in their recent equivalents, these facies are distributed in fossil reef complexes as a function of the biological and hydrodynamic zonation of the depositional environment (Fig. 6.50).

Main types of deposits. Various types of platform carbonate deposits have been described from formations ranging from the Cenozoic to the Precambrian (cf. Miall 1984; Sellwood, in Reading 1986). These include:

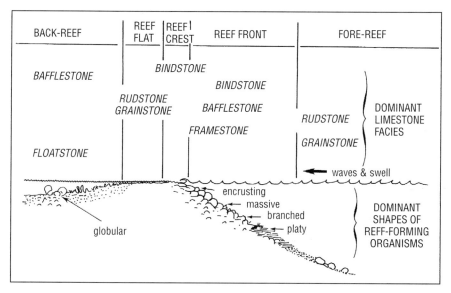

Fig. 6.50. Schematic section through a fossil reef: general arrangement, shape of building organisms, and main limestone facies (after James, in Walker 1984); *grainstone* sandy grains in contact with each other, without carbonate matrix (see Table 1.5)

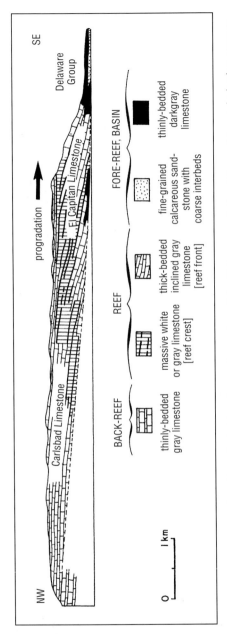

Fig. 6.51. Transverse section through the Permian reef complex of the Delaware and Guadeloupe Mountain basins, western Texas (after King, in Miall 1984)

1. Soft deposits only cemented during diagenesis and developed during periods of high sea-level when vast areas of the continents were covered by shallow seas. This was the case for certain Ordovician limestones of North America and the Cretaceous chalks of Western Europe. This type of deposits, which sometimes is continuous over several tens of kilometers along strike, was sparsely developed during the recent stages of fairly low sea-level.

2. Structureless benthic deposits extending from emerged stable zones. These are particularly well developed in the upper Devonian of Alberta in Canada, where they compare closely with those forming at present on the Bahama Bank and along the coast of Yucatan.

3. Well-structured reefs on the outer shelf of ancient seas in comparatively stable environments. They constitute systems of barrier or fringing reefs, like in the Permian complex in the Delaware basin of Texas (Fig. 6.51). This important massif, extending laterally over several kilometers, exhibits a well-developed facies zonation from the back-reef through the reef itself and then into the fore-reef, with a strong seaward inclination of up to 30° formed during a progradation over the shelf for some 20 km. Modern reefs, like the Great Barrier Reef of Australia, the carbonate domes of the Florida Keys, and the Tulear Reef of Madagascar (Sect. 6.3.3.2) appear also to belong to this type.

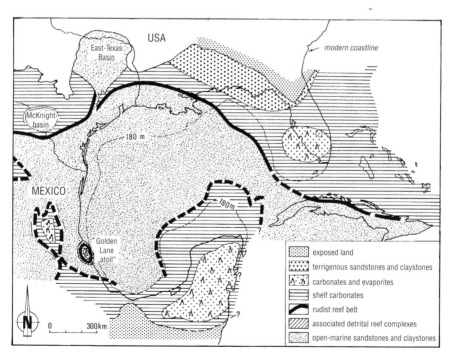

Fig. 6.52. Sedimentary environments in the lower Cretaceous rudist barrier reef of the Gulf of Mexico (after Bryant, Meyerhoff et al., and Vernon; Sellwood, in Reading 1986)

4. Partially built complexes situated along the edge of emerging areas during phases of uplift. A good example is represented by the upper Jurassic (Portlandian-Purbeckian) of southern England where the reefs are restricted to algal massifs and ridges (stromatolites) developing within coastal environments of variable energy.

5. Partially built carbonates in subsiding environments. These are located at present mainly around the margins of certain volcanic archipelagos of the Pacific and Indian Oceans. These types of carbonate complexes were widely distributed in the past, especially during the Paleozoic and the Cretaceous. The lower Cretaceous of the Tethys in the Gulf of Mexico and its margins show the presence of an irregular-shaped barrier made up of rudists. This belt was limited in the landward direction by terrigenous deposits, precipitated carbonates and evaporites, and shelf carbonates. In the seaward direction it borders on open-marine limestones and claystones (Fig. 6.52). This arrangement which shows little prograding tendency, is marked by a strong subsidence leading to vertical growth of the reef barrier. In the sheltered back-reef environment, a sedimentary fill consisting of carbonates and aggrading terrigenous deposits is developed.

7 Open-Marine Environments

7.1 Submarine Relief and Sedimentation

7.1.1 Introduction

Open-marine sedimentation is controlled by a variety of factors, the most important of which are submarine relief, movement of water masses, climate, sources of terrigenous and volcanogenic material, biologic productivity, processes of dissolution, and phenomena of resedimentation. The influence of these factors has been dealt with in detail by authors such as Lisitzin (1972), Hsü and Jenkyns (1974), Kennett (1982), Seibold and Berger (1982), Jenkyns (in Reading 1986), and Rupke (in Reading 1986). Additional information is furnished by specialized treatizes like Heezen and Hollister (1971), Reineck and Singh (1980), Warme et al. (1981), Galloway and Hobday (1983), and Miall (1984). We shall restrict ourselves to a resume of the main conditions of sedimentation as determined by the critical factor of submarine relief, the characteristics of which are confined solely to the marine environment.

7.1.2 The Continental Margins

The continental margins occupy some 20% of the submarine regions of the earth (Figs. 7.1 and 7.2), and comprise three morphological units (cf. Vanney 1977; Boillot 1979; Blanc 1982).

1. The *shelf* constitutes the extension of the relief of emerged coastlines below the sea. Its depth extends from sea-level down to between 100–250 m with frequent slight inclinations of 0.07° on average. Its width varies from about a kilometer, as along the Maritime Alps of southern France and west of Corsica, to some 1000 km in northwestern Europe and the China Sea. The recent sedimentary cover of the shelves is made up of highly diverse material ranging from boulders to clay as a result of the variability of littoral and coastal environments and the fluctuations of sea-level during the upper Quaternary (cf. Sects. 6.2 and 6.3).

228   Open-Marine Environments

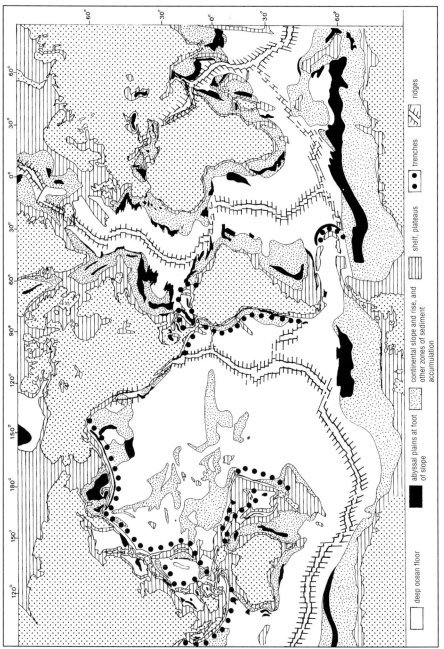

Fig. 7.1. Main elements of submarine topography (after Heezen and Hollister 1971)

2. The *continental slope* extends down to between 2000–4000 m. It usually starts at a break in the slope of the shelf, and its inclination, although frequently in the range of 3°–5°, can locally attain values of 10° or more. The slopes are somewhat hilly where they run out on submarine plateaus like the Blake Plateau off Florida, or the Rockall Plateau south of Iceland. They are frequently dissected by submarine *canyons* resulting mainly from subaerial erosion by running waters (e.g., off the Rhone and the Mississippi deltas) which took place during Plio-Quaternary phases of eustatic regression, or prior to the subsidence of the modern continental margins. There are, however, canyons such as the Scripps canyon off California, which lack any apparent connection to any exposed fluvial system. They result either from submarine erosion of loose or poorly lithified sediment by turbidity currents or slumps, or along grabens, that is, tectonic trenches associated with fault systems. These canyons represent preferential pathways for coarser-grained sediments which here are able to traverse the continental slope. The slope deposits themselves are usually fine-grained and can be distinguished from shelf sediments or those of coastal or deltaic environments by the widespread occurrence of gravitational movements. These frequently result in coarser mean grain sizes at the base of sedimentary successions or in depressions in zones of dispersal. The upper or proximal portion of the slope is typically dominated by sedimentary reworking, by erosion and the incision of channels, and by the transit of sand on its way to the ocean floor. In contrast to this, the distal part of the slope is more a zone of sediment accumulation or agradation, partly through decantation and partly through resedimentation. It shows a gradient decrease towards the continental rise or the abyssal plains. Locally, the transitional zone between shelf and slope is the preferred place for the formation of ferruginous granules (glauconites) or phosphates.

3. The *continental rise* is only developed at the foot of certain passive continental margins subjected to active sedimentation (cf. Sect. 7.2.1). It consists of accumulations of sediment in the form of shallow cones or lobes, depending on whether the sediment transport takes place more at right angles to the coast by gravity currents or parallel to it by contour currents. Its sediments which are dominated by silty-sandy oozes, can cover vast areas like the outer Blake ridge. This extends for over 1100 km off the east coast of the USA (Fig. 7.8). Note that the continental slope and rise represent important potential zones of hydrocarbon accumulation trapped within turbiditic sands intercalated in hemipelagic oozes, and in structural traps along growth faults and diapirs.

7.1.3 Abyssal Plains

The abyssal plains, making up some 80% of the oceanic domain, comprise basins, ridges, and trenches (Figs. 7.1 and 7.2). The *oceanic basins* consist of:

1. The abyssal plains situated at the foot of the continental margins at depths of between 4000–5000 m in mid-oceanic basins where rates of

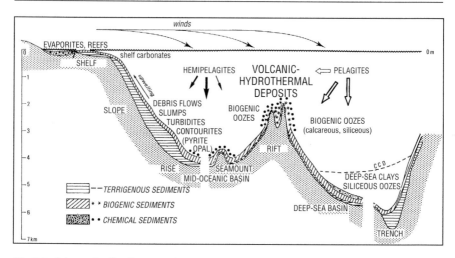

Fig. 7.2. Schematic distribution of terrigenous, biogenic, and chemical oceanic sediments (vertical scale highly exaggerated); *CCD* carbonate compensation depth

sedimentation are rather high (above 1 cm/1000 years), and dominated by hemipelagic oozes with abundant biogenic material which increases in proportion away from the shelf.

2. The deeper intra-oceanic basins (5000–6000 m) which are better protected from terrigenous supply. These are marked by very slow sedimentation of the order of millimeter per 1000 years, by intense dissolution of carbonates, and to a large extent by pelagic sediments (siliceous oozes, deep-sea clays). Reducing sediments containing sulfides and opaline tests can form over oxygen-depleted bottoms on both the plains and basins. The oceanic basins are interrupted by hills, seamounts, and even complete archipelagos which are mostly of volcanic origin. However, they sometimes result from tectonic movements or diapirism. On some of these elevations abraded summits are developed as a result of subaerial erosion prior to thermal subsidence of the crust: the so-called guyots. Sedimentation on the summits of seamounts is weak, resulting mainly from decantation, and locally it may be missing, especially in the presence of active bottom currents. In contrast to this, sedimentation on the flanks of seamounts can be considerable due to gravitational reworking of sediment in the direction of the adjoining basins.

The *oceanic ridges* are immense submarine mountain chains with a central valley, accompanied by adjacent fractured crests and plateaus. They extend for more than 60 000 km along zones of accretion between oceanic plates. The average water depth over them is about 2500 m, but locally they even breach the sea-level as in Iceland and the Azores. Together with the recent aseismic rises and the fossil ridges they represent the major obstacles inhibiting marine circulation and sediment transport. Only supply by decantation, mostly of planktonic tests and, to a lesser degree, by aeolian transport, can add to the products of volcanism and hydrothermal processes. Sedimentation along the ridge axes is usually rather weak, becoming more important on the flanks where processes of resedimentation add to the products of decantation.

The *oceanic trenches* which are zones of subduction situated along active oceanic margins of the Pacific type or along island arcs like Japan, Indonesia, and the Marianas, represent remarkable sediment traps. Their mean depth is about 6000 m, but locally they may reach 11 000 m as in the Mariana Trench. Terrigenous sediments supplied by decantation and resedimentation together with volcaniclastic deposits accumulate here and at the same time fill the basins, to form a rise like the southern portion of the trenches off South America. However, where the supply is low, the rate of sedimentation can be moderate or totally lacking compared to the rate of subduction. This is the case where arid climates prevail on adjacent continental areas, where there are no large river systems debouching into the sea (e.g., in the east-Pacific trenches north of Valparaiso/Chile), or where the trench is sheltered from the continent by an intervening morphological barrier like the Mariana trench east of the Marianas. In the oceanic trenches, biogenic deposits are of secondary importance and chemical deposits are rare. Locally, reducing environments may develop as a result of a weak exchange of the bottom waters as in the Puerto Rico trench in the Caribbean.

7.2 Dynamics of Deep-Water Sedimentation

7.2.1 Continental Margins

7.2.1.1 Conditions for Sedimentation

Along the continental margins the conditions for sedimentation are largely controlled by the global tectonic context:

1. Passive margins of the Atlantic type present by far the most diverse conditions. These are: (i) decantation of hemipelagites, laminated deposits resulting from mixed suspensions derived by reworking from shelf muds; (ii) slumps, debris flows, and turbidity currents along the axes of submarine canyons, or triggered by excess loading, earthquakes, or liquefaction etc., along the top of the continental slope, (iii) accumulation of relatively ordered and mature detrital fan complexes extending from large subaerial or submarine drainage systems; (iv) transport and deposition of particulate matter by seasonally appearing masses of waters of differeing densities (phenomena of cascading due to rise or downward plunging of waters); and (v) deep thermohaline circulation controlling contour currents which are particularly active along the western margins of the oceans. In general, the fluvial supply of material to the sea tends to take different pathways, depending on the grain sizes (Moore 1969). The sand is initially subjected to littoral drift along the

coast until it encounters the head of a canyon whence it is carried to the open sea. Silt and clay accumulate in coastal muds, settling out in the seaward direction where they are then remobilized by various gravitational currents.

2. Active margins of the Pacific type are frequently characterized by irregular, immature, and heterogeneous deposits with a notable volcanogenic component. The sediments of back-arc basins like those of Japan and Indonesia resemble those of passive margins along continental coastlines rather than those of active margins along island arcs.

3. Margins with transform zones like the Gulf of California exhibit a horst-and-graben topography controlling a rather irregular pattern of sedimentation such as detrital fans, organic accumulations, and condensed deposits.

7.2.1.2 Mechanisms of Resedimentation

Depending on the particular location, the Atlantic margin of the USA presents several distinct sedimentary processes:

1. The continental slope, extending from 120–140 m to 2000 m depth at an inclination of 3°–10° is deeply incised by canyons of Pleistocene age, notably the Hatteras and Hudson canyons. Its sediments are dominated by a thick prograding succession of Tertiary to Quaternary age, supplied with hemipelagites from the adjacent continent (Keller et al., in Doyle and Pilkey 1979). The sand content decreases from above 30% on the distal shelf to below 10% at the foot of the slope. Exceptions occur along the axes of the active Hudson, Hydrographer, and Veatch canyons off the New England coast which contain turbidites. Slumps occur over wide areas supplying material as far as the Sohm and Hatteras abyssal plains. Bottom currents with velocities of up to 0.7 m/s are observed occasionally along the slope. Their effects, however, as yet are poorly known.

2. The rather extensive continental rise is dominated by sedimentation of silty-clayey material derived from the north as a result of a deep thermohaline circulation system. The Western Boundary undercurrent originates in the Norwegian Sea and extends down to the coast of South America (Fig. 7.3). Deep currents locally attain velocities of up to 0.5 m/s, resulting in deposits of silts, intercalated clays and fine-grained sands: the so-called contourites of Heezen et al. (1966). These sediments even out bottom irregularities, and locally attain a thickness of up to 2 km. The numerous sediment drifts identified along the foot of the North Atlantic continental margin appear to result from such contour currents. These extend from south of Iceland down to the northern Antilles on the western side and off the Straits of Gibraltar on the eastern side (Fig. 7.3; Stow and Piper 1984). Giant mud waves of 10–40 m in height, with wave lengths of 1–5 km have been observed on certain of these drifts (Roberts and Kidd 1979). Their detailed structure and mode of formation are still poorly known.

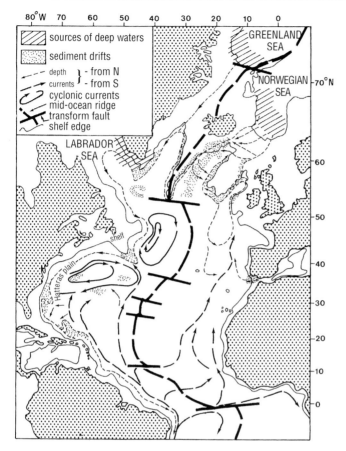

Fig. 7.3. Deep oceanic circulation in the North Atlantic with main recent sediment drifts (after Stow and Holbrook, in Stow and Piper 1984)

7.2.1.3 Deep-Sea Fans

Submarine sediment fans are dominated by detrital supply and extend from the top of the continental slope down to the adjacent abyssal plains. They exhibit certain similarities with subaerial alluvial fans (cf. Sect. 5.4). The dispersion of the sediments in them is controlled by a system comparable to the migrating channels bordered by levees and separated from each other by interchannel flats. These systems are rather complex, exhibiting a general downstream ordering of sedimentary megasequences which permit the distinction of proximal and distal units. The deep-sea fans, in which sediments resulting from turbidity currents predominate, develop from a localized sedimentary source at the top of the margin. This source is funneled through a fluvial system debouching into a system of canyons, such as in the fans of the Ganges, Niger, Mississippi, and the Rhone. It can also consist of littoral or coastal alluvials without an individual fluvial source, such as the east coast of the USA, California, and the Cape Ferret fan in the Bay of Biscaye, or of glacial material like the Pleistocene St. Lawrence fan off the Canadian east coast.

Based on the horizontal extent, three main types of sedimentary units rich in turbidites have been identified by Stow et al. (1983/1984) at the base of continental margins:

1. Elongate fans form as an extension of a single principal sediment source, frequently represented by a river. They usually possess a sole main supply channel which changes downstream in a complex way. The sediments comprise facies of rather diverse extent, mostly dominated by clay, silt, and fine-grained sand. Fans of this type are found off the mouths of the Ganges, Indus, Mississippi, Amazon, St. Lawrence, and the Rhone. From fossil sequences such complexes, albeit of a more moderate size, have been described from the Tertiary of Piemont/NW Italy and the North Sea (cf. Bouma et al. 1985).

2. Radial fans develop in a semi-concentric pattern at the mouth of canyons or isolated submarine channels which frequently show no relationship to fluvial systems. They are marked by a more restricted detrital supply than in the case of the elongate fans described above. Sandy deposits are abundant, sometimes even prevailing over oozes. This situation is developed in the La Jolla, San Lucas, and Redondo fans off the west coast of North America, and in the Tertiary complexes of the North Sea (Frigg, Magnus), in California, and along the margins of the Alps and the Pyrenees.

3. Slope-apron deposits are derived from a more dispersed upstream source area without a distinct supply channel. They are made up of sediments rich in sand with minor gravel, more or less associated with oozes, due to the low to medium rates of sedimentation. They are developed over a number of sections of the recent oceanic margins away from the zones of direct fluvial discharge, and probably were equally widespread in the geological past.

As a result of the diversity of the factors responsible for the origin of deep-sea fans, there are various intermediate stages between the three types described above (Fig. 7.4). The most important influences appear to be exerted by tectonic activity and fluctuations in sea-level. The best-developed fans and slope sediments are situated mainly along active margins and in their vicinity where tectonic activity leads to intense accumulation of material. During periods of transgression, sedimentation in the fans is rather weak as continental erosion tends to diminish when the sea-level rises. The opposite situation is developed during tectonic quiescence and low stands of sea-level.

A number of *models for deep-sea fans* have been proposed as a result of regional studies (cf. Walker 1984; Bouma et al. 1985). Comparison of known recent fans with those reconstructed on land leads to even more questions, as the former are mostly known by their shape and their surficial sediments whereas the latter have to be reconstructed from vertical sections. Despite their apparent homogeneity, the deep-sea fans can be subdivided in the downstream direction into three different stages, characterized by different

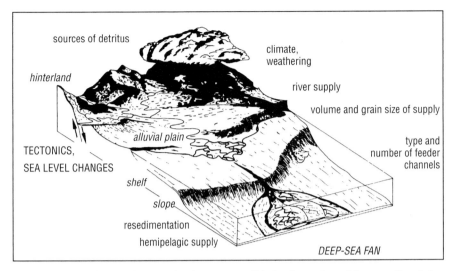

Fig. 7.4. Sketch of the various mechanisms responsible for formation of deep-sea fans (after Stow et al. 1983/1984)

major facies. Figure 7.5 depicts the morphological and sedimentary model proposed by Walker (1978), which is especially applicable to the radial or only moderately elongated fans:

1. The upper fan is situated at the mouth of the feeder channel which usually is a canyon with steeply inclined walls and steep longitudinal gradient. The upper fan is mainly comprised by a channel with varying sinuosity which is strongly confined by its banks or levees, which can extend above it by several dozens of meters and even up to 75 m, like on the Rhone fan (Bellaiche et al. 1981). The length of the channels varies between a few and several hundreds of kilometers, with their width ranging from several hundred meters to more than 25 km. The channel sediments are particularly diversified in fans developed along rising mountain ranges. In their upper section, these channels are made up of unordered sequences of debris flows and beds of gravels and boulders. These have been subjected mainly to erosive reworking, and they grade laterally into slumps (Fig. 7.5). In the downstream direction some degree of ordering becomes apparent, increasing towards the distal zones as follows: (a) sand and gravel with inverse and then normal graded-bedding due to friction along the channel bed and then by a waning of the transporting current; (b) sediments with normal graded-bedding when the depth increases and the energy of the particle flow is diminished; and eventually (c) the development of stratification. This evolution of the channel deposits constitutes an excellent criterion for the longitudinal polarity of the complex. The banks exhibit more homogeneous sedimentation of fine-grained turbidites with divisions (d) and (e) of Bouma (cf. Chap. 3), temporarily interrupted by coarser-grained sediments from spill-overs or from channel migration. The sequences

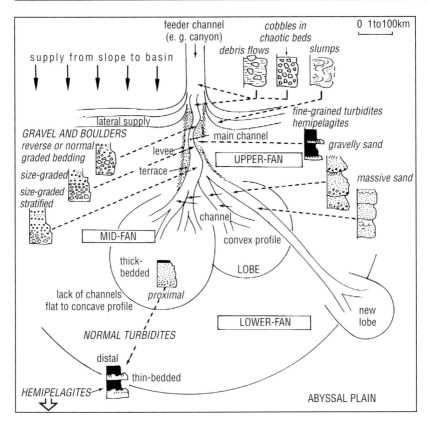

Fig. 7.5. Facies distribution and basic sedimentary sequences in a typical deep-sea fan characterized by abundant sand supply and unconfined flow (after Walker 1978). Scale of a lobe is in the order of several dozens of kilometers

deposited on the upper fan, with thicknesses between 10 m and more than 90 m, show a upward-fining tendency (Fig. 7.6) when sedimentary progradation leads to the filling of the main feeder channel by outer-shelf deposits.

2. The mid-fan is characterized by *sedimentary lobes* which become gradually more inactive as they advance downstream. These lobes are supplied with sediment from the main channel which spreads out into a network of branching, sinuous, and sometimes braided minor channels (Fig. 7.5). The average gradient decreases markedly in zones of strong sedimentation, leading mostly to a convex surface as on the Amazon, Indus, and Rhone fans. The course of the channels can be deflected by the Coriolis force (Cremer 1985). Near zones of intense tectonic activity, their floor is occupied by sequences of gravel and sand, or coarse-grained sand showing scour casts and lenticular bedding (Walker 1978). The sides of the channels and their low relief banks are made up of graded-bedded sands covered by clayey silts, representing Bouma divisions (d) and (e) of proximal turbidites. This arrangement is

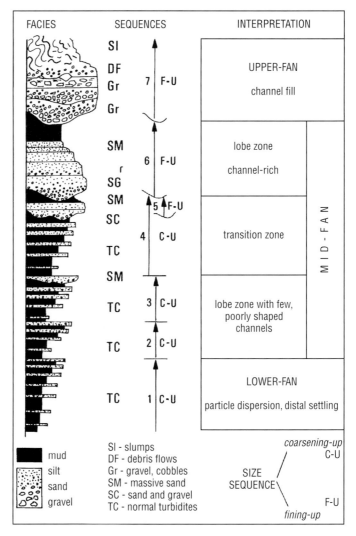

Fig. 7.6. Schematic sedimentary sequence in a sand-rich deep-sea fan in an actively prograding environment (after Walker 1978)

changing continuously due to: filling, abandonment, and lateral migration of channels in fining-upward sequences (Fig. 7.6); incision into inter-channel zones of sedimentation by migrating new channels in coarsening-upward sequences; coverage and development of new lobes during the course of progradation, or dissection of previous lobes by a transgressive regime (Fig. 7.7 B). The sandy lobe sequences frequently make up packages 10–50 m in thickness. In modern detrital fans rich in silty-clayey material, like those of the Mississippi and the Rhone, channel networks are rather poorly developed.

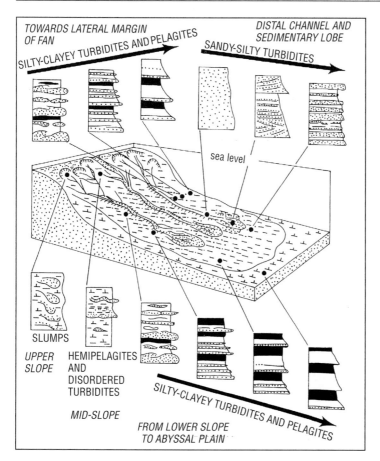

Fig. 7.7. Schematic distribution of sedimentary facies in a large deep-sea fan (after Stow and Piper 1984). Note the systematic distribution of the various types of fine-grained turbidites parallel and at right angle to the fan axis, and the presence of hemipelagites and pelagites at various levels within the fan

Debris flows with fine-grained matrix and slumps are frequently developed, whereas sandy fills are rare.

3. The lower fan, representing the zone below the detrital cone proper extends widely over the foot of the continental slope and can even advance onto the adjacent abyssal plain, as, for example, below the Indus and Ganges cones. Its slightly concave surface shows only a weak gradient, and may be cut by rare, poorly defined channels. Material supplied by gravity processes is limited to *distal turbidites* with Bouma divisions (c), (e), and sometimes (b), alternating with hemipelagites which increase in thickness and frequency in the downflow direction (Fig. 7.6). Progradation leads to covering of the more or less rhythmic sediments of the lower fan by the more sandy sediments of the mid-fan, with a tendency to channel formation and upward-coarsening sequences.

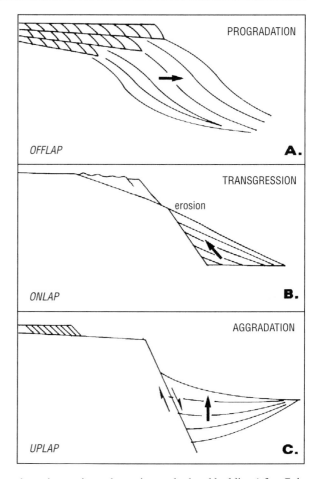

Fig. 7.8. Types of detrital fans along the continental margins, and related bedding (after Galloway and Hobday 1983). The *thick arrows* indicate the sense of progressing sedimentation

This model of distribution of sedimentary facies down the detrital fans is frequently difficult to reconstruct in ancient sequences, especially under poor outcrop conditions. Detailed studies show furthermore that, as in modern complexes, and in the absence of notable tectonic and eustatic constraints, local factors lead to a facies distribution which is more complex than foreseen in the models (cf. Galloway and Hobday 1983; Stow and Piper 1984; Walker 1984; Bouma et al. 1985; Reading 1986).

Numerous modern detrital fans present a prograding regime with offlap deposits or descending wedges particularly pronounced along the margins of stable continents in zones of strong fluvial supply, and during the Quaternary regressive periods of low sea-levels (Fig. 7.8). This is the case for the fans of the Rhone, Nile, Ganges, Niger, and Mississippi.

Table 7.1. Extent of some deep-sea fans (after Moore et al. 1978)

Deep-sea fan	Depth at top (m)	Depth at base (m)	Length (km)	Width (km)	Area (10^3 km^2)	Volume (10^3 km^3)	Maximum thickness (km)	Thickness along edge (km)	Age (10^6 years)
Amazone	1,500	4,800	520	600	215	710	14	1	8
Ganges	1,600	5,000	3,000	1,000	3,000	10,000	12.5	1	50
Mississippi	1,200	3,400	350	600	170	85	?	12	3.5–6

The large, laterally extremely extensive and very thick prograding fans are evidence of a rather long and complex geological history (Table 7.1). Fans subjected to strong destruction by incision of canyons in the upper part, and with onlapping deposits in ascending wedges, correspond to transgressive regimes during which the sediments of the upper slope and the external shelf become reworked, and participate in a landward progression of sedimentary material. A good example of this is the Tertiary of the Gulf of Mexico (in Galloway and Hobday 1983). When developed on aggrading fans as onlap deposits, they grow in thickness, especially along strongly faulted and subsiding margins, and their seaward extent is moderate, as along the Californian coast or along the margin of certain ancient flysch basins.

7.2.2 Open-Marine Basins

7.2.2.1 Abyssal Plains

Abyssal plains occupy portions of the oceanic basins situated off passive continental margins, and also occur on the inner side of certain back-arc basins like the Bering, Japan, or China Seas (Fig. 7.1). They represent the most distal environments in which sediments of continental origin are deposited by gravitational processes. Despite their very weak depth gradients (in the order of 1:1000), they frequently are made up of turbidite beds derived either from the adjacent margins, seamounts or ridges, and from fine-grained suspension clouds from the shelves. The abyssal Hatteras plain, off the southeast coast of the USA, constitutes a good example of these mixed sedimentary environments subjected to sedimentation by decantation and resedimentation (Fig. 7.9). The layer attains a thickness of 350 m in the southern part which receives silty-sandy turbidites derived from the Hudson and Hatteras canyons with average thicknesses of 0.5–2 m. In contrast to this, a thin sediment layer covers the higher portions of the ocean bed, as found by geophysical prospecting. Certain individual turbidites coming from the margins, the so-called black-shell turbidites of Pilkey et al. (1980), spread out over almost 1000 km in a north-south direction with a width of up to 400 km. Other, less well-studied turbidites appear to be of a more local origin. The major turbidity cur-

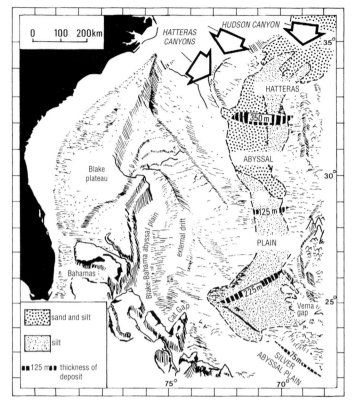

Fig. 7.9. Distribution of zones of sandy-silty turbidites on the Hatteras abyssal plain (after Rupke, in Reading 1986)

rents, which are initially perpendicular to the continental margin, start to become gradually more parallel to it, once they have exited from the canyon. Note that semi-closed seas like the Mediterranean and the Gulf of Mexico, are the places of active hemipelagic and at the same time gravitational sedimentation, the sources of which can be rather variable for a given locality (Rupke, in Reading 1986).

7.2.2.2 Intraoceanic Basins

Relatively protected against sedimentation from gravity or density flows, intraoceanic basins are poorly known from the point of view of sedimentary dynamics. Decantation here frequently plays the dominant role. However, the numerous topographic highs represented by the oceanic ridges, aseismic ridges, plateaus, seamounts, and trenches, together with the bottom currents which increase with topographic narrowing and around areas of positive relief, allow a degree of sedimentary reworking which can be rather important locally. There is a diversity of evidence for these processes: turbidites and other resediments intersected by drill cores in the basins at the foot of carbonate platforms (Purser 1983; McBreath and James, in Walker 1984); point measurements of strong, multidirectional currents along the axes of ridges and in transform zones; current structures at all depths affecting various types of oceanic oozes and also the fields covered by polymetallic or Mn-nodules (Heezen and Hollister 1971); and mechanical deformation of certain brown-red pelagic oozes where current structures are combined with the effects of bioturbation (Pautot and Hoffert 1984).

7.2.2.3 Deep-Sea Trenches

The trenches constitute preferred places for sedimentary reworking resulting from the deepening of the ocean floor due to subduction, by the formation of an accretionary prism, and by the uplift of the adjacent continental regions. The nature of the gravitational transport mechanisms varies considerably with the rate of terrigenous supply, the morphology of the adjacent continent and of the oceanic floor, and with the rate of subduction (cf. Bally 1983). For example, the trench situated east of the Japan isles which is experiencing moderate fluvial supply but bordered by steep slopes, receives the products of mass sliding or slumping issuing from the internal wall. The slides can be considered as "tectonic erosion" processes triggered by the subduction. This results in a tendency for sedimentary filling through transport at right angles to the trench. The sediments thus accumulating lead to a morphology made up of narrow basins which then are subducted together with the oceanic crust and its sediment cover (Cadet, Kobayashi et al. 1985). In contrast to this, the Nankai trench south of Japan receives abundant turbidites issuing from the fluvial Fuyi basin which becomes rejuvenated and uplifted by the collision of the Izu and Honshu regions (Taira and Niitsuma 1985). Deposition takes place parallel to the axis of the trench in the form of a very thick fill leveling out the irregularities of the oceanic substrate. The turbidites fill practically the whole trench and are integrated in the accretionary prism together with the underlying hemipelagites as subduction progresses northward under Japan (Le Pichon 1985). Perched sedimentary basins can develop in depressions on the accretionary prism and become deformed under the influence of subduction together with the turbidites on the bottom of the trench (Fig. 7.10). Subduction-induced folds and faults in the accretionary prism can lead to the liquefaction of water-saturated sediments and to the expulsion of methaniferous fluids in seepages which are prime localities for the development of bacterial colonies and benthic abyssal faunas consisting of giant bivalves, worms, etc. Under the influence of subduction, the sedimentary successions actively filling the trenches generally tend to become gradually coarser as one passes from the pelagic beds at the base to distal products of fill and then to the proximal ones. The study of these successions has to be combined with detailed structural analyses if one wants to identify such deposits of fossil trenches. So far, however, only few examples have been described (cf. Reading 1986; Leeder 1982).

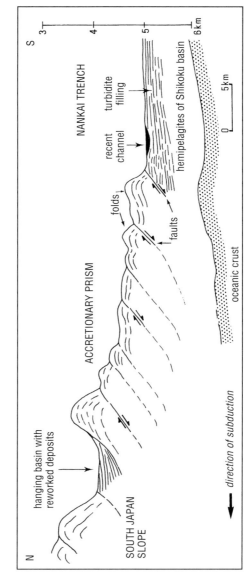

Fig. 7.10. Seismic profile across the southern Nankai trench, south of the Japan Isles (R./V.J. Charcot, Franco-Japanese cruise Kaiko I, leg 1: Le Pichon, Iiyama et al. 1985)

7.3 Main Deep-Sea Sediments

7.3.1 Carbonate Oozes

The deposition of pelagic carbonate oozes is favored by moderate water depths of less than 4000 m, by warm surface waters, and restricted supply of terrigenous material (Fig. 7.11). These conditions are developed in particular in the central parts of the Atlantic at low to medium latitudes (Fig. 7.12, Table 7.2). Due to the intense dissolution of material as a result of its great depth and the aggressive nature of its bottom waters, the Pacific Ocean is relatively poor in carbonate oozes. They are mostly restricted to the active ridge zones, to aseismic ridges, and to plateaus and archipelagos. The Indian Ocean represents an intermediate situation between the two other large oceans. The mean rate of deposition of carbonate-rich oozes is in the range of 3 cm/1000 years (Table 7.3), the variations depending especially on the planktonic productivity, the size of the sedimenting tests, and on the importance of the terrigenous supply. The nature of the carbonate sequences varies with time mainly as a function of the water temperature. This permits a number of paleoclimatological and paleo-oceanographic applications (e.g., Fig. 7.13).

There are three types of carbonate oozes which are developed on ocean floors of medium depth, mixed partly with other deposits:

1. Foraminiferal oozes which are rich in sand-sized tests, represent the most common facies, particularly in zones of high productivity. With a theoretical settling velocity of about 2 cm/s, the foraminiferal tests reach the bottom within several weeks, even when marine waters are turbulent.

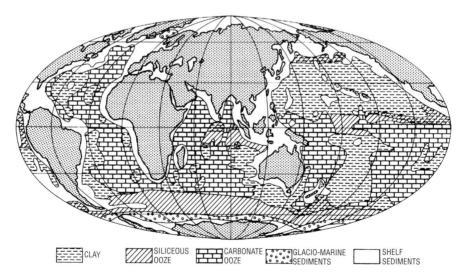

Fig. 7.11. Distribution of main types of recent deep-sea sediments (after Berger, in Kennett 1982)

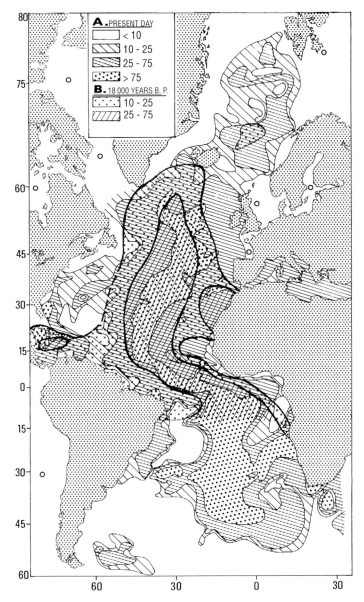

Fig. 7.12. Distribution of CaCO$_3$ in recent sediments (*A*) of the Atlantic Ocean compared to last glacial maximum (18 000 years BP) in the North Atlantic (*B*) (after Biscaye et al. 1976); ○ = area not considered

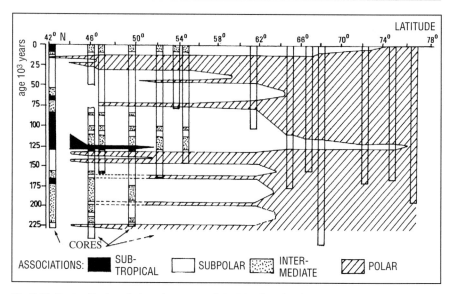

Fig. 7.13. Latitudinal variations in waters of the North Atlantic and the Norway Sea during the middle and upper Quaternary, as concluded from planktonic foraminifera and coccolithes (after Kellogg, in Cline and Hays 1976)

Table 7.2. Distribution of main types of pelagic sediments (after Kennett 1982)

Sediments	Atlantic	Pacific	Indian ocean	Total (%)
Foraminiferal and nannofossil oozes	65	36	54	46.5
Pteropod oozes	2	0.1	–	0.5
Diatom oozes	7	10	20	12
Radiolaria oozes	–	5	0.5	3
Clay	26	48.9	25	38
Relative size of ocean (%)	23	53	24	

2. Coccolithe oozes, or nannofossil oozes, comprise the calcareous remains of coccolithophorids and of the now-extinct discoasters. Due to their particular structure, calcareous nannofossils are more resistant to dissolution than foraminifera. They survive in sediments of slightly greater depths and they are found more in zones of lower productivity. The small mean size of the coccoliths (about 10 µm) leads to settling velocities and sedimentation rates which are much lower than those of foraminifera. The slow settling time of about 100 years to the ocean floor is frequently compensated by the mass descent of coccoliths with fecal pellets produced by zooplankton which contain up to 10^5 coccoliths per fecal pellet of 50–250 µm diameter. The coccoliths incorporated in these pellets are particularly well protected against dissolution in sea-water and easily descend to ocean floors of 5000 m depth within less than two months.

3. Pteropod oozes consist of debris of planktonic gastropods, the size of which ranges from 0.05 to 10 mm. The aragonitic nature of pteropod shells tends to favor an active dissolution at depth or a transformation to calcite. Pteropod oozes occur rather locally at depths lower than 3000 m (Table 7.2), especially in temperate-warm basins (Mediterranean, Atlantic, Pacific).

7.3.2 Siliceous Oozes

The deposition of recent biosiliceous oozes which are dominated by diatoms and to a lesser extent by radiolarians (Table 7.2), is directly linked to zones of high planktonic productivity (Fig. 7.14). The siliceous sediments mostly occur in the form of latitudinal belts (Fig. 7.11):

1. The main, 900–2000-km wide belt occurs around Antarctica and its rim of glaciogenic sediments. It results especially from upwellings, the rise of nutrient-rich bottom waters, caused by the violent storms originating on the Antarctic continent. Its northern limit coincides with the Antarctic convergence, the zone of downward plunge of the surface waters. This belt contains some 75% of the silica fixed in recent oceanic sediments. Siliceous tests, very rich in diatoms, can account for up to 70% of the sedimentary mass. The sedimentation rate of the siliceous tests, which is rather low due to their low density, can be increased by the formation of fecal pellets which furthermore inhibits the strong dissolution normally taking place in the surface waters. The accumulation of the light diatomaceous shells takes place preferentially in the oceanic basins.

2. A second belt, less widespread but equally rich in diatoms, occupies the northern Pacific and the adjoining Bering and Ochotsk Seas. The concentration of the siliceous tests here is strongly diluted by terrigenous material.

3. An equatorial belt is well developed in the Pacific and locally in the Indian Ocean. Its position is determined by the upwelling of nutrient-rich deep waters under the influence of the dominantly westward winds. The siliceous sediments, in which radiolaria predominate, are mixed with carbonate sediments also linked to high planktonic productivity. The rate of sedimentation of the radiolaria with 4–5 cm/1000 years is considerably less than the 10–20 cm/1000 years recorded for foraminifera west of the East-Pacific Rise. In the shallower zones over the oceanic ridges the sediments are enriched in carbonate particles, whereas in the deeper zones away from the equatorial belt they are enriched in pelagic clay (cf. Sect. 7.3.3).

More localized concentrations of sediments rich in biogenic silica are encountered along the west coasts of the major continents, like off southern and western Africa, Peru, and California, resulting from coastal upwelling induced by the trades. These upwellings of cold waters rich in nitrogen and phosphorus, although seasonal in nature, frequently lead to high rates of productivity and of sedimentation. In the Gulf of California, the annual rates

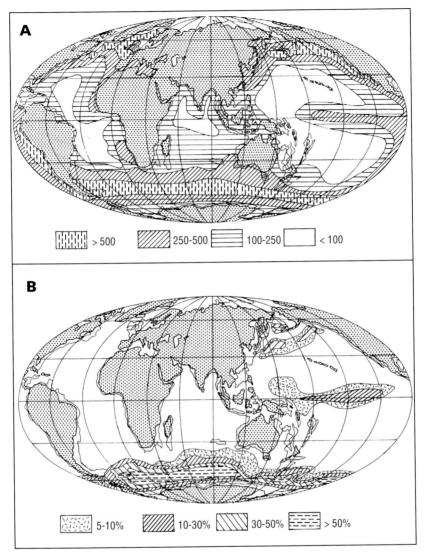

Fig. 7.14 A, B. Rates of extraction of dissolved silica by phytoplankton (**A**) from surface waters (SiO_2 in 10^{-2} g/a) and content of biogenic silica in surface sediments (**B**), corrected to 0% $CaCO_3$ (after Calvert 1974)

are as high as 0.08 g/cm² against values of about 0.2 g/cm² in the biosiliceous sediment belt surrounding Antarctica. The siliceous tests are associated especially with carbonates and phosphates. In general, the evolution of these belts and their concentrations of the siliceous organisms with time appear to be controlled mainly by planktonic productivity and climate (Jenkyns, in Reading 1986).

7.3.3 Clays and Argillaceous Oozes

Clay minerals are present in all oceanic sediments, frequently constituting a notable to dominant fraction (Fig. 7.11, Table 7.2). The most widespread ones are illite, smectite, and kaolinite. Chlorite and random mixed-layer minerals are common, whereas other clay minerals, such as vermiculite, palygorskite, sepiolite, and regular mixed-layer minerals are less common and only of local importance.

Sedimentary compounds associated with the clay minerals assist in characterizing the oceanic sediments: silt and sand in terrigenous argillaceous oozes, carbonate and siliceous tests in hemipelagic oozes, organic matter and sulfides in sapropelic black shales, opal and zeolites in certain pelagic oozes, metal oxides in brown-red deep-sea oozes, etc. All possible transitions between the three main types of oceanic sediments, viz. carbonate oozes, siliceous oozes, and clay may occur. The rate of sedimentation of clays is low in the very deep open oceanic environments (<1–10 mm/1000 years; Table 7.3), a fact favoring the formation of polymetallic or manganese nodules and micronodules, as well as the relative enrichment of micrometeorites. The rate increases considerably in zones of strong terrigenous supply, e.g., to 5–15 cm/1000 years in the hemipelogic oozes of the Mediterranean, independent of the various phenomena of resedimentation.

Most of the recent oceanic clays are of continental derivation. This is proven by a number of observations: parallelism between the clay cover of terrestrial surface formations (rocks, weathering blankets, soils) and marine sediments of the same latitude (cf. Chamley 1989); increase of the content of minerals derived from intense surface hydrolysis (kaolinite) in sediments of

Table 7.3. Rate of sedimentation of recent pelagic sediments (after Berger 1974)

Sediments	Region	Rate in mm/1000 years
Carbonate oozes	North Atlantic 40–50°N	30–60
	Equatorial Atlantic	20–40
	Caribbean	28
	Equatorial Pacific	30
	East Pacific Rise 0–20°S	20–40
	East Pacific Rise 30°S	3–10
	East Pacific Rise 40–50°S	10–60
Siliceous oozes	Equatorial Pacific	2–5
	Antarctica, Indian Ocean side	2–10
	North and Equatorial Atlantic	2–7
	South Atlantic	2–3
	Extreme North Pacific	10–15
Argillaceous oozes	North Pacific, central portion	1–2
	Northern tropical Pacific	0–1
Clays, silts	Off Californian coast	50–2,000
	Southwest Atlantic, Ceara plain	200

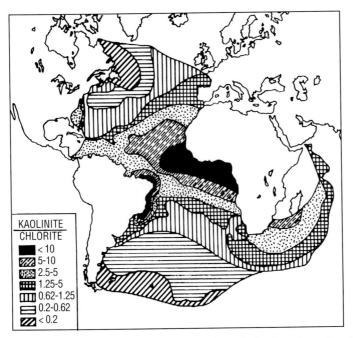

Fig. 7.15. Variations in the kaolinite-chlorite ratio in the <2 µm fraction of surface sediments in the Atlantic (after Biscaye, in Kennett 1982). The values correspond to the areas under the XRD-peaks for kaolinite (3.58 Å) and chlorite (3.54 Å) resp.

warm, humid regions, in contrast to the increase of minerals derived from physical weathering (chlorite) in sediments of cold regions (Fig. 7.15); relative abundance of illitic micas in zones of the world oceans under strongly terrigenous influence (high latitudes, northern hemisphere, large rivers draining shield areas; E-W belt in the northern Pacific subjected to strong aeolian supply, etc.; Fig. 7.16); absence of notable chemical modifications during transport, variations of the assemblages with time as a result of changes in continental climate of alluvial sources, of tectonic activity, etc. (Chamley 1971, 1986; Singer 1984). In general, it can be stated that terrigenous illite and chlorite are derived mainly from plutonic and metamorphic rocks, and kaolinite from well-drained soils in warm climates. Smectites originate in soils of temperate to humid regions, and palygorskites in pedogenic carbonate crusts or ancient evaporitic sediments, whereas the random mixed-layer minerals are derived from little-weathered soils and regoliths (Griffin et al. 1968; cf. Chamley 1989). These various minerals can have experienced more than one cycle of erosion and deposition, and reach the recent oceans via the erosion of ancient sedimentary rocks, like the kaolinite of the northern oceans which is inherited from peri-arctic Mesozoic sediments.

Smectites, resulting from a number of genetic processes and forming preferentially in volcanic rocks highly susceptible to weathering, exhibit a distribution pattern less dependent on the latitudinal zonation than those of the other clay minerals. The predominant portion of sedimentary smectites resulting from volcanic rocks are derived from the reworking of weathering products accumulated on dry land (Sect. 7.3.6.1). The volcanogenic smectites originating in the truly marine environment are mostly only abundant within or close to the "parental" lavas or pyroclastics. They can also form through chemical exchange along the surface of extremely slowly accumulating sediments like the above-mentioned abyssal oozes. Fibrous clays are either of detrital derivation, being dispersed within other clay minerals, or of authigenic origin, tied to volcano-hydrothermal activity, or located within concretions or within individual layers. Note that the phenomena of sedimentary inheritance after the upper Jurassic, in combination with the frequently localized or discrete nature of the processes of oceanic clay diagenesis, facilitated the use of clay associations in numerous paleogeographic reconstructions (Chamley 1981, 1989; Robert 1982; Singer 1984).

In the oceanic environment, *mineralogical sorting* is generally only weak, as the essential processes of differential sedimentation have already taken place in the littoral and shelf environments (cf. Chap. 6). The very small size of the clay minerals, frequently below 1 µm, leads to their vast dispersal, permitting a characterization of their alluvial origin and of the respective water bodies far removed from the sources. Smectites are particularly suited to long-distance transport because of their small size, their flocculated nature, and their stability against further flocculation. They can, however, also be deposited before the other clay minerals. This appears to be the case in the Panama basin where smectites occurring in organo-mineral aggregates and fecal pellets in the surface waters, are rapidly deposited on the continental slope. In contrast to this, illites and chlorites settling out in deeper waters are dispersed farther out into the sea. Furthermore, sedimentary reworking can displace sediments of the continental margin rich in smectites farther out to greater depths (Honjo et al. 1982).

The brown-red oozes of deeper water, the so-called "deep-sea red clays" are particularly widespread in the Pacific and, to a lesser extent, in the Indian Ocean. They are frequently depleted in biosiliceous, carbonate, and sand-sized debris; their color is caused by various manganese and iron oxides. They show a great diversity in composition and origin. Certain clays are essentially authigenic in origin, as in the central East Pacific where various types of smectites are forming together with zeolites (philippsite), metalliferous concretions or aggregates etc., depending on whether the environment of deposition is influenced by volcanism or by the presence of residual planktonic tests consisting of silica or carbonate (Hoffert 1980). Other "red clays" with various types of mineral associations are essentially terrigenous in origin, having been brought in by wind action, or resulting from decantation as in the deepest portion of the North Pacific during the upper Cretaceous (e.g., Lenôtre et al. 1985; Fig. 7.16). A continental derivation of marine clays is indicated: by the latitudinal correspondence in clay mineral distribution between oceans and adjacent continents; by the absolute age of the clay minerals deposited in the

252 Open-Marine Environments

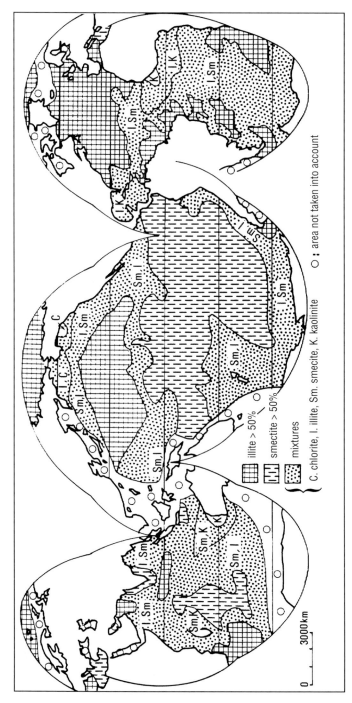

Fig. 7.16. Main clay minerals in the <2 μm fraction of recent oceanic sediments (after Berger 1974). Areas indicated as "mixed" designate the absence of a content of one mineral of above 50%; the main minerals then are mentioned in decreasing order of abundance

oceans which reflects that of the adjoining geological environments; by the abundance of aeolian quartz and other detrital minerals in the clays; and by the absence of notable environment-controlled influences on the clay particles. All intermediate stages, in which a precise characterization is very difficult, can exist between authigenic and terrigenous-derived deep-sea red clays. Like other oceanic sediments, the pelagic deep-sea clays are susceptible to reworking by bottom currents or gravitational processes.

7.3.4 Polymetallic or Manganese Nodules

Characteristics: The polymetallic nodules, frequently referred to as manganese nodules, are mainly associated with deep-sea brown oozes (cf. Glasby 1977; Marching 1981; Académie des Sciences 1984; Hoffert 1985). These oxidic nodule-shaped or crust-like concentrations are found at various depths and in different environments which, as a common feature, experience only slow or episodic sedimentation (Fig. 7.17). The nodules, consisting of a black core and a brown rim, have mean diameters of 1–10 cm and are made up mainly of oxides of Mn and Fe, together with various other elements of economic interest, notably Ni, Cu, and Co (Table 7.4). Nodules smaller than 1 cm are referred to as micro-nodules. The estimated mineable reserves of economically interesting elements in the nodules are considerable: 220–690×10^6 t Ni, 40–227×10^6 t Co, 0.190–0.626×10^9 t Cu and 4.5–16.3×10^9 t Mn (Acad. Sci. 1984). Iron and manganese occur in the nodules in the form of oxides and poorly crystallized silicates, together with other authigenic or detrital minerals such as clays, silicates, and carbonates, associated with planktonic tests. They are usually made up of concentric layers around a nucleus consisting of different types of material such as rock fragments and fish teeth. The nodules lying on the ocean floor frequently have an upper, smooth portion, enriched in Fe and Co in contact with water, and a lower, irregular portion enriched in Fe and Mn in contact with the sediment. This indicates that there are notable chemical exchanges between the nodules and their environment. Furthermore, there are significant differences in their chemical composition depending on their geographical location, thereby emphasizing the important participation of ocean waters in their formation. Those in the Pacific are richer on average in Mn, No, Co, and Cu, whereas those in the Atlantic are richer in Fe. Nodules of the Indian Ocean occupy an intermediate position.

The concentration of manganese nodules is particularly high at the water–sediment interface over brown clays within a depth range of several hundred meters immediately below the carbonate compensation depth which occurs at about 4500 m. Significant bottom currents can lead to displacement of the nodules, thereby explaining their spheroidal shape and concentric structure, and their sometimes broken-up nature.

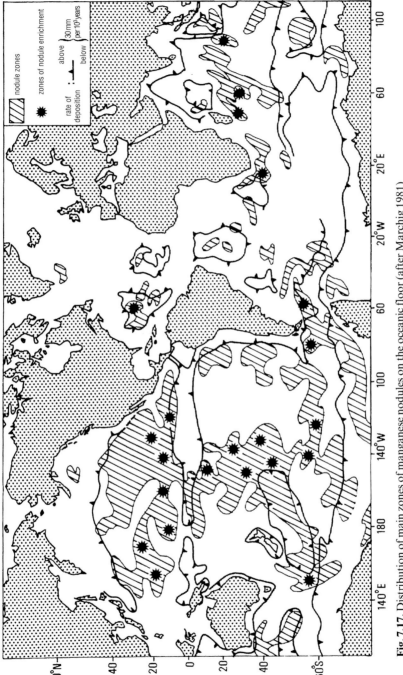

Fig. 7.17. Distribution of main zones of manganese nodules on the oceanic floor (after Marchig 1981)

Table 7.4. Average physico-chemical data on manganese nodules (Académie des Sciences 1984)

Mn	29.5%	Al	1.12%
Fe	6.34%	K	0.8%
Si	4%	Ti	0.3%
Na	2.92%	Co	0.25%
Mg	2.88%	S	0.23%
Ca	1.44%	P	0.14%
Ni	1.40%	Zn	0.14%
Cu	1.16%	Mo	0.06%

Predominating manganate: todorokite
Density: about 2
Specific surface area: 100–200 m^2/g
Porosity: 30–50%
Concentration in zones of economic interest: 7 kg/m^2 (locally > 20 kg/m^2)
Depth of maximum concentration: 4,000–5,200 m

The mean growth rate of the nodules is extremely low, in the range of only a few millimeters per million years. Thus it is considerably lower than the several mm/10^6 years of the sediments upon which they formed and are now preserved. This paradox can be explained in a number of ways: dissolution of nodules embedded in a reducing environment; relocation of the nodules, or erosion of the surrounding sediment by bottom currents; and migration to the sediment surface as a result of bioturbation or differences in density. Note that the growth of the nodules can be rather irregular, with long periods of interruption, as shown by radiometric dating and biostratigraphic studies of successive layers. The nodules furthermore bear evidence of displacement along the sea bed, such as rounded or broken features, erosion marks, grain-size sorting, and preferred orientation of their long axes.

The formation of manganese nodules necessitates a concentration of their elements in, and precipitation from, sea-water or sediment in the oxidizing environment of the red clays subjected to active bottom currents. Manganese, iron, and the other constituting elements are derived from distant continental weathering crusts, or from hydrothermal activity and alteration of subaqueous volcanic rocks. The relative importance of the continental and oceanic contribution to the water masses, and the degree of remobilization of already precipitated metals from the host sediments, are responsible for the differences in composition of the nodules from different geographic locations. The mechanisms fixing the elements around the nucleus are still poorly known and understood, but various ideas have been put forward:

1. direct precipitation under oxidizing conditions (acceleration of bottom currents, intrusion of cold waters from high latitudes);

2. indirect precipitation due to biological, and particularly bacterial activity;

3. concentration in tests of organisms;

4. concentration in surface sediments due to dissolution in deeper, more reducing layers and recycling by upward migration of the dissolved material;

5. absorption by subamorphous manganese aggregates which exhibit a considerable specific surface area with a strong negative charge.

The formation of the nodules took place throughout the Tertiary over the brown oozes. Part of the concretions observed now on the ocean floor results from slow accumulations, from movements of sediment, and from particularly low rates of sedimentation. Manganese nodules have also been formed during other geological periods, such as those in the oceanic red clays of the Upper Cretaceous of the Indonesian island of Timor (Jenkyns, in Reading 1986).

7.3.5 Metalliferous and Hydrothermal Sediments

Sediments enriched in iron and/or manganese, together with various other elements like Cu, Cr, Pb, and Zn, are typically associated with oceanic ridges (Rona 1984). Such sediments, with iron contents locally above 20%, frequently cover the surfaces of basalts derived from ocean floor spreading. They are moved laterally from the ridge by further spreading and then covered by normal sediments (Fig. 7.21). Such *basal metalliferous sediments* are encountered in boreholes intersecting the interface between oceanic crust and overlying sediments. They are also developed in certain recent rift zones like the Red Sea graben. Iron tends to become precipitated in abundance first, followed by the less concentrated and less readily oxidized manganese which deposited preferentially farther away from the spreading axis and at a lower rate. This difference in the rate of precipitation of oceanic crust-derived elements facilitates the recognition of proximal, iron-rich volcano-hydrothermal, and more distal, Mn-rich deposition in ancient sedimentary sequences. Furthermore the relation between oceanic volcanic and terrigenous influences can be traced from the distribution and composition of these sediments (e.g., Debrabant and Chamley 1982; Fig. 7.24).

The resulting hydrothermal deposits occur in a number of different forms such as bedded metalliferous sediments, crusts over basalts or normal sediments, mound-shaped accumulations and fragile chimneys of 10-m height or more. These are the so-called black or white "smokers", from which emanate hot waters charged with suspensions of mineral precipitates during productive phases. The hydrothermal production is frequently of only a temporary nature. Thermal waters escaping from fissures in the oceanic basalts are sometimes rather hot (up to 375° C) and frequently enriched in minerals, especially sulphur. This element is utilized for the growth of bacterial colonies which synthesize organic matter. The organic matter is then taken up by communities of exceptionally large organisms such as bivalves and worms which are not dependent on light. They are similar to the communities forming around points of discharge of methane-bearing fluids on accretionary wedges (Sect. 7.2.2.3).

The main minerals occurring in hydrothermal metalliferous deposits are oxides with variable stages of hydration like goethite and limonite, various silicates (clay minerals, quartz, opal), calcium sulfates and sulfides. Among the clay minerals, ferriferous smectites (nontronite) are dominant. Their formation in the Bauer basin west of South America is favored in shallow depressions by the presence of metallic oxides and opal, and in the higher-lying zones by calcite (Cole 1985). The sulfides, rich in Zn, Cu, Fe and locally associated with native silver, can be of economic interest. Recent hydrothermal accumulations are known from a number of restricted sites occurring predominantly along ridges or zones of transform faulting, as for example: In the East Pacific (Juan de Fuca Rise between 13°–21° N and the Galapagos Rise at 20° S); in the Atlantic (Famous Zone near the Azores and the Romanche Deep); the Red Sea, and the Carlsberg ridge in the Indian Ocean. In certain sub-basins of the Red Sea, such as the Atlantis II Deep, the hydrothermal deposits, with a thickness of up to 20 m contain economic reserves of Zn and Cu, and are overlain by a 200-m thick zone of up to 60° C hot brines.

The formation of hydrothermal deposits is linked to the presence, along the axis of oceanic ridges, of young oceanic crust not yet covered by sediment

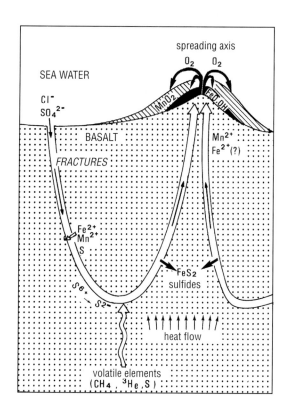

Fig. 7.18. Schematic formation of hydrothermal deposits on ridge axes (after Bonatti 1975; see text for explanation)

and rendered permeable by intense fracturing of basalts during cooling. The cold oxidizing and alkaline marine bottom waters penetrate downward along these fractures, heating up in the process and taking up elements derived from the alteration of basalts (Fe, Mn, Zn, Cu, Ba, Ca, Si, etc.) or migrating upward from deeper magma chambers (^{3}He, S, CH_4). In this way, the waters become hot, acid, and reducing, and enriched in various elements. They then reascend to the surface (Fig. 7.18), resulting in progressive cooling and differential precipitation of various minerals which is mainly controlled by the duration of ascent. Important factors here are the extent of fracturing and the interaction with cold water derived from the open marine environment. Other types of mineralization can result from the interaction of basalts with seawater especially where volcanic sills intrude into unconsolidated sediments. Locally, processes of secondary metamorphism take place through heating of the sediments and the formation of secondary minerals such as smectites, corrensite (smectite-chlorite regular mixed-layer), chlorite, locally palygorskite; zeolites (analcime, mordenite, wairakite, locally clinoptilolite); quartz, epidote, and apatite, etc. Typical examples are found in the North Pacific and the Marianas basin (Desprairies and Jehanno 1983).

The various types of hydrothermal deposits depend mainly on the geodynamic situation of the locality in question (cf. Hoffert 1985). In ocean ridges in presently forming basins, like the Gulf of California or the Red Sea, they are underlain by shallow magma chambers at a depth of about 1 km or slightly more. The endogene-derived precipitations are interfingering with exogenic deposits of a terrigenous, biogenic, or chemical nature such as clays, carbonate or siliceous oozes, organic matter, or evaporites. This leads to rather varied hydrothermal deposits consisting of oxides, silicates, sulfates, and sulfides. Ridges in older oceanic basins usually harbor hydrothermal deposits of a less diversified nature. Where the spreading rate is below about 4 cm/year, as in the Atlantic, the magma chambers are deeper and situated below 5 km. Also, the fracturing network is dense and complex, and the solutions advance to deeper levels and at slow rates. This results in stockwork deposits especially rich in iron, accumulating within the basalts. Surface deposits are less frequent and, where present, richer in manganese. In basalts of rapidly expanding ridges of the Pacific type, the solutions proceed only to shallower depths of 1–2 km and ascend more rapidly. Deposits around localized points of discharge are relatively abundant, and are rich in sulfides and manganese instead of iron.

Various types of economically interesting fossil hydrothermal deposits are known from Cyprus, Oman, New Caledonia, and elsewhere. They are associated with ophiolites, that is, ancient oceanic rocks brought to the surface through intense tectonic thrusting.

7.3.6 Other Sediments

7.3.6.1 Volcaniclastic Rocks

Sediments containing material resulting from volcanic eruptions are particularly abundant close to active island arcs and emerging portions of oceanic ridges, at the foot of which they can form thick accumulations, as in the West and North Pacific, the Indonesian archipelago, and off Iceland (cf. Fisher and Schmincke 1984). Volcanogenic sediments, consisting of fragments known as *tephra,* comprise materials deposited from wind transport and those deposited by dense particulate or fluid mass flows, such as ignimbrites. Wind-deposited volcaniclastics are by far the most frequent types in recent and fossil oceanic sequences. They make up the pyroclastics, which are distributed around eruptive centers under the direct control of the prevailing winds (Fig. 7.19). The pyroclastics consist of particles of highly contrasting size, from bombs to lapilli (63–2 mm) and ash (2–0.6 mm), and volcanic dust. The majority of oceanic pyroclastic sediments range in size from fine-grained sand to clay, and are made up of porous or non-porous glass, together with local feldspars, pyroxenes, and other heavy minerals. Individual volcaniclastic sources can be reconstructed from volcanic particles dispersed in normal sediments which result from weathering and erosion of subaerial volcanoes in the same way as those of other geological or pedogenic formations.

Depending on the extent of their dispersion, volcaniclastic sediments are present in three main types:

1. Local ash falls, deposited mostly within several tens of kilometers from the volcanic vent. These are made up of a wide range of particle sizes. Their

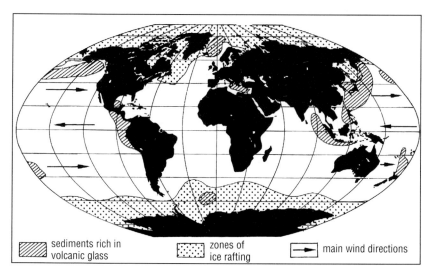

Fig. 7.19. Main zones of oceanic sediments enriched in volcanic glass (after Ninkovich, in Kennett 1982) and zones with glacio-marine sediments (after Heezen and Hollister 1971)

heavy mineral content and mean grain size decrease away from the source. An upward decrease of grain size is frequently observed in beds made up of volcanic glass. Superimposed onto this trend, there may be an increase in grain size (inverse graded-bedding) due to the presence of highly porous pumice of high floatability. Pumice fragments can be transported on the water surface over long distances before sinking or being deposited along beaches. They are thus subjected to the directions of oceanic surface currents. This is the case for the floating pumice of the Aeolian islands north of Sicily, which is transported by the north-Mediterranean current up to the Gulf of Genoa and even as far as the Provence beaches of Southern France.

2. Falls of tropospheric derivation result from volcanic particles ejected to heights of 5–12 km and transported over distances of several thousands of kilometers within periods of less than 1 month. They are the products of volcanic explosions which occur frequently when considered on a geological time scale, i.e., about one every 10 years. Size sorting during transport is pronounced, with grains of 20–100 mm being deposited over distances less than 1000 km, and grains of 1–20 µm being deposited over greater distances. The sources, however, can generally still be reconstructed.

3. Falls of global extent comprise fine-grained volcanic dust (0.3–1 µm) ejected to great altitudes. These particles have residence time of several years and are disseminated around the whole globe parallel to the equator. These are deposited on land and sea, independently of the location of the original source. Their small particle size and their dispersion throughout the various sediments make their recognition difficult.

The study of the distribution of volcaniclastic beds in recent and fossil sedimentary sequences facilitates stratigraphic correlations (Fisher and Schmincke 1984; Lajoie, in Walker 1984). Detailed petrographic studies and radiometric dating frequently allow the identification of the source volcanoes, the reconstruction of the dispersal pattern, and the dating of the enclosing normal sediments (tephrochronology). As an example, the two major upper Quaternary pyroclastic horizons in the eastern Mediterranean may be mentioned (Fig. 7.20). The first one, erupted about 25 000 years B.P. from the island of Ischia off Naples, covers a large portion of the central and eastern basin. It comprises alkali-trachytic glass accompanied by aegirine-augite, biotite, apatite, sphene, sanidine, and minor hornblende and plagioclose. The second one, erupted about 3450–3500 years B.P. from a collapsed caldera in the Santorin archipelago of the Aegean Sea is restricted to a smaller portion of the eastern Mediterranean. Its deposition was contemporaneous with, and at least in part responsible for, the disappearance of the Minoan culture of northern Crete. It consists of dacitic glass with hypersthene, plagioclase and other trace minerals.

The *alteration of volcanic particles* after deposition is poorly known as the alteration products, dominated by smectites and zeolites, can completely replace the original grains which are thus difficult to identify. Weathering of volcanic materials is fairly rapid in subaerial moisture-rich environments

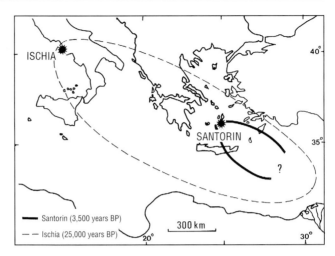

Fig. 7.20. Dispersion of pyroclastics ejected in the Mediterranean Sea by the Ischia volcano about 25 000 years B.P. and by the Santorin caldera about 3500 years ago (after Keller and Ninkovich 1972)

whence the alteration products are dispersed frequently into the seas thereby masking the clay products of marine in-situ alteration (cf. Sect. 7.3.3). The volcaniclastic particles are more susceptible to submarine alteration as their size decreases and their porosity increases. This is due to their increase in specific surface area. The presence of biosiliceous debris made up of unordered opal from diatoms and radiolarians also makes reactive silica available for the construction of silicate lattices (Chamley 1971). The highly porous pyroclastic materials locally allow access of hydrothermal fluids which tend to accelerate the process of halmyrolysis or subaqueous weathering. From the 90° East Ridge of the Indian Ocean, Besse et al. (1981) described a 400-m thick sequence of vitreous ash and lapilli. Their constituent materials are altered, probably by hot waters, to Mg-smectite, analcime, and clinoptilolite at the base, and ferriferous smectite, clinoptilolite, mordenite, and then to philippsite to the top.

7.3.6.2 Non-Volcanogenic Aeolian Sediments

The source regions of marine aeolian deposits are characterized by arid or desert climates, subjected to strong winds of constant direction. Atmospheric turbulence systems have to remain constant over considerable time spans to permit aerosol particles to stay in suspension over long distances. Rain over the oceans is generally responsible for the deposition of the particles. Areas of relatively important aeolian deposition are developed off pericontinental deserts like the Sahara, Namibia, and Australia, and over the Pacific (cf. Fig. 7.19). Estimates of the material issuing from the Sahara are in the range of $60-200 \times 10^6$ t/year. These aeolian dusts in general are very fine-grained and characterized by reddish iron oxide coatings. They are rich in quartz, but frequently contain other minerals such as calcite, feldspar, and clay, as well as rather fragile species like palygorskite (attapulgite) and sepiolite (Bain and Tait 1977). Furthermore, particles of organic derivation are transferred from land to sea during sand or dust storms. This is the case for the phytolites, opal particles secreted by plants, fresh-water diatoms (mostly from equatorial regions), and for spores and pollens (mainly from mid-latitude regions).

During aeolian transport, sorting processes may develop which, with increasing distance, lead to a selection of more resistant, smaller particles, e.g., quartz in the size range of 1–20 µm. The preferred deposition of aeolian quartz in the sea off the large arid continental regions permits interesting reconstructions of position and character of the climatic zones during the Quaternary. Periods of increased glaciation are marked in the sediments off the Sahara by an increase of the quartz supply, by a coarser grain size, and by a wider latitudinal spread of the zone of aeolian deposition. The increase of the seaward winds off the large deserts furthermore led to increased upwelling of nutrient-rich abyssal waters during the glaciations. This resulted in characteristic deposits indicative of strong planktonic productivity along the continental margins (cf. Sect. 7.3.2).

7.3.6.3 Glacio-Marine Sediments

Sediment particles resulting from the melting of ice are encountered over a large part of the high-latitude ocean floors (Fig. 7.19). They are made up of materials of highly diverse nature and size, ranging from giant dropstones to clays (cf. Sect. 5.1.2). The Antarctic Sea is characterized by the important role played in sedimentation by the icebergs issuing from the ice shelves. The icebergs are carried northward by the circumpolar current, eventually melting down to negligible size before reaching the Antarctic convergence. They transport annually some $35-50 \times 10^9$ t of sediment (Lisitzin 1972). The belt of glaciogenic sediments in the Southern Ocean between the Antarctic continent and the belt of diatomaceous oozes (Sect. 7.3.2), is characterized by heterogeneous particles decreasing in mean size to the north (in Kennett 1982). The Arctic region which, save for Greenland, does not possess an extensive ice cap, is characterized by widespread outcrops of land during the annual snow melt. These furnish abundant fine-grained glaciogenic sedimentary material to the oceans, resulting mainly from permafrost soils, and transported by rivers during sudden ice flow discharges. During the Quaternary, the glacially influenced marine zones extended to lower latitudes, as can be reconstructed from the detailed petrographic, mineralogical, and morphoscopic study of marine drill cores. The glaciomarine sedimentation during the middle and upper Quaternary in general was more important in the northern than in the southern hemisphere. Of the total glaciogenic sediments of this period, 88% occur in the northern hemisphere (North Atlantic 62%, Norway Sea 12%, North Pacific 8%, Arctic Ocean 6%), against only 12% around Antarctica (Ruddiman 1977). This results mainly from the fact that the Antarctic inland ice represents a polar desert in which the thickness of the ice cover together with the absence of active drainage systems protects the rocks from notable subaerial erosion.

7.3.7 Global Tectonics vs Deep-Sea Sedimentation

Formation of basaltic crust along the oceanic ridges, together with the expansion of the sea floor, is the reason for a number of global sedimentary phenomena characterizing the geological history of the marine sedimentation and the water masses covering the sediments. The sediments deposited directly on the crust become increasingly younger closer to the ridge and their thickness increases in the direction of the continental margins. The subsidence resulting from crustal cooling, coupled with the increasing sediment load, leads to a deepening of the sea floor as spreading continues. This causes all marine zones subjected to sufficient subsidence, to pass at a certain point in time of the oceanic history, through the carbonate compensation depth (CCD) which is situated at a depth of 4000–5000 m. This explains why at a given point of the sea floor, the planktonic carbonate oozes deposited on the basalts after an intervening layer of more or less metalliferous hydrothermal sediments, generally grade upwards into argillaceous deep-sea oozes or into siliceous oozes (Fig. 7.21). The reappearance of carbonate sediments can result from a temporary lowering of the CCD because of an increase in the reservoir of dissolved carbonate due to lower productivity, a period of climatic cooling, a eustatic drop in sea-level, passage of the given section of the lithospheric plate through a more productive climatic zone, or an increase in the circulation of oxygenated abyssal waters. The study of these sedimentary successions is the field of oceanic *plate stratigraphy* which has numerous applications in recent and ancient oceans (cf. Kennett 1982; Jenkyns, in Reading

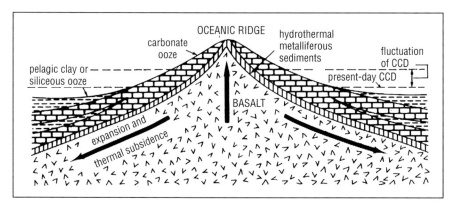

Fig. 7.21. Sedimentary successions developing during sea-floor spreading and subsidence of oceanic crust (after Berger and Winterer 1974)

1986). In the following, only a short introduction into these problems is presented.

The *Pacific plate* which has been permanent since the Mesozoic and was sheltered from variations in terrigenous supply by its bordering trenches, is migrating northwestward under conditions of continuous deepening. The underlying sediments were made up essentially of carbonate oozes when the depth of the floor exceeded the CCD, siliceous oozes when depth and planktonic productivity were high, carbonate and siliceous oozes under shallow depths and strong productivity, and argillaceous deep-sea oozes within highly productive ocean zones (Lancelot 1978). The equatorial zone represents the main region of biogenic carbonate and silica production, and thus results in a strong increase of the rate of deposition when the plate passes through this belt. The post-upper Cretaceous sedimentary successions encountered in the various boreholes of the Deep Sea Drilling Project show considerable lithological differences, depending on whether the oceanic crust formed north of the equatorial zone, on it, or to the south of it (Fig. 7.22). In general, the sediments deposited to the north of this zone represent a simple thin succession of limestones and red clays, with poorly developed layers of siliceous oozes (radiolarians) making their appearance when, at the point of the establishment of the CCD, the zone of high productivity was fairly close (Fig. 7.22A). In contrast to this, the sediments deposited south of the equatorial zone present a complete succession from carbonate oozes, to

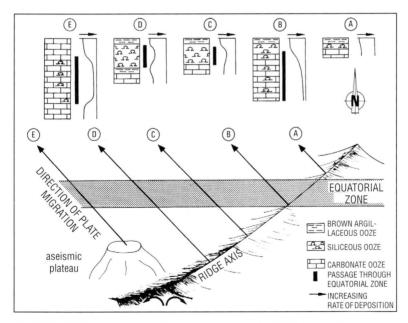

Fig. 7.22. Different types of sedimentary sequences in Pacific boreholes as a function of the place of formation of the oceanic crust in relation to the equatorial belt (after Lancelot 1978; see text for explanation)

siliceous, and then argillaceous oozes, sometimes with a recurrence of argillaceous oozes when the depth of deposition was already considerable prior to the establishment of the zone of higher productivity (Fig. 7.22 D). The aseismic rises and plateaus situated to the west, such as the Shatsky, Ontong-Java, Hess and Magellan zones, have not yet experienced major subsidence and thus have not passed beyond the CCD. Because of this, they are covered by continuous limestone deposits which constitute excellent markers for fluctuations in the rate of sedimentation resulting from the passage of the Pacific plate through the equatorial zone. This expresses itself in larger rates of sedimentation and the occurrence of radiolarians (Fig. 7.22 E). The systematic recording of the lithological successions in oceanic boreholes facilitate the reconstruction of the sedimentary, oceanographic, and climatic history of the Pacific Ocean (Lancelot 1978). This approach requires an exact knowledge of the age of the respective sediments (biostratigraphy) and of the fluctuations with time in direction and velocity of the plate migration. The poles of rotation can for example be reconstructed by the "hot-spot method".

The *Atlantic Ocean* has spread and deepened since the mid-Mesozoic in a direction mostly parallel to the equatorial zone. Thus, its sediments are less dependent on variations of productivity than on terrigenous supply, the regional tectonic situation, and on the movement of the water masses. The rather complex sedimentary sequences deposited since the Jurassic are composed of such diverse materials as evaporites and dolomitic limestones, reduced or oxidized argillaceous oozes, biocalcareous and biosiliceous oozes, zeolitic clays and hemipelagic oozes. The continental influence is pronounced throughout the entire history of the Atlantic, due mainly to the following reasons: an initially rather narrow ocean constantly bordered by close land areas; virtual absence of sedimentary traps (trenches); abundant fluvial supply; and tectonic instability of the margins during spreading. The Atlantic region furthermore was particularly sensitive to paleo-oceanographic changes. This occurred initially during the growth of the basins in the upper Cretaceous when a deep circulation system was established, and then during the Cenozoic when worldwide cooling led to vertical exchanges of water in the form of thermohaline circulation (cf. Kennett 1982). The recording of the various lithofacies intersected by oceanic boreholes, in combination with a study of the curves of subsidence furnished by geophysical investigations, and with the identification of sedimentary gaps from geophysical and biostratigraphic data, permits reconstruction of the sedimentary and paleo-oceanographic history of the various basins of the Atlantic (e.g., McCoy and Zimmerman 1977). The relatively simple geodynamic history of the South Atlantic represents a particularly fruitful field of study (e.g., Melguen et al. 1978; Fig. 7.23). Three major periods have been identified here:

1. Passage from an intracontinental rift to a confined narrow marine basin open to the south and north between the Valanginian and Santonian (127–82 m.y.). Characteristic sediments are: terrigenous sands, evaporites, dolomitic limestones, then argillaceous oozes, carbonate and organic oozes.

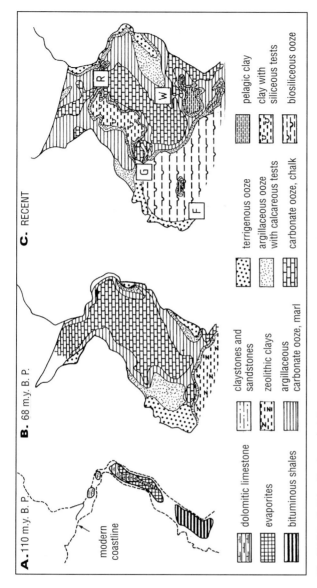

Fig. 7.23 A–C. Distribution of sedimentary facies in the South Atlantic during the Albian-Aptian (**A**), Maastrichtian (**B**), and Recent (**C**) (after Melguen et al. 1978). Submarine barriers: *R* Romanche; *G* Rio Grande; *W* Walvis; *F* Falkland

2. Progressive north- and southward opening, accompanied by deepening and rupture of the morphological barriers of the Romanche, the Walvis-Rio Grande, and of the Falkland-Agulhas ridges between the Campanian and the mid-Oligocene (82–35 m.y.). Characteristic sediments are: multicolored argillaceous oozes and carbonate oozes, expressing fluctuations of the CCD.

3. Development of the modern situation of thermohaline circulation and latitudinal climatic zonation since the Oligocene, characterized by hemipelagic and carbonate oozes.

The detailed study of the sedimentary constituents extends our knowledge of the history of modern and ancient oceans and their borderlands through lithological studies. This is especially the case for micropaleontologic data which, together with analyses of oxygen and carbon isotopes, supply extensive information on: the establishment of connections between the various oceanic basins; eustatic sea-level fluctuations; surface and bottom circulation; the migration of climatic factors; variations in organic productivity; and the relative importance of terrestrial and cosmic phenomena (cf. Kennett 1982). This is equally true for the organic matter. The origin of its components and their stage of evolution permit the recognition of stratigraphic markers, and furnish information on the confinement of the environment, on the relative importance of terrestrial and marine influences, and on the hydrocarbon potential of the sediments (e.g., de Graciansky et al. 1982).

Data on the *mineralogical and geochemical stratigraphy* from the scale of sediment smear slides to the analysis of individual grains with the aid of the microprobe, also assist in a better understanding in a number of fields. These include mechanisms of transition between oceanic and continental influences, paleogeographic conditions on the continents adjacent to the sedimentary basins, the effects of local or global tectonics on sedimentation, presence or absence of morphologic barriers, the nature of the detrital sources, and characterization of the ancient oceanic water masses:

1. In the North Atlantic, the mineralogic and chemical sequences reflect the competition between oceanic volcanism and terrigenous supply, tectonic instability of the margins alternating with continental peneplanation, the redox conditions over the ocean floors (e.g., Fig. 7.24), modifications in global climate, and migration of the lithospheric plates through different climatic zones (Chamley and Debrabant 1984).

2. In the intra-plate basins of the North Pacific, global tectonics is manifested by subsidence and the latitudinal migration of the oceanic crust as these are sheltered from the effects of changes on the continental margins by the belt of trenches surrounding them. In the Marianas basin, for example, the Cretaceous sedimentary succession consists of an ordered sequence reflecting geodynamic or geochemical events (Chamley et al. 1985): proximal volcanism and local supply from emerged archipelagos, then early marine diagenesis associated with lithospheric subsidence and increasing distance to volcanic sources, and eventually disappearance of the submarine barriers, diversifica-

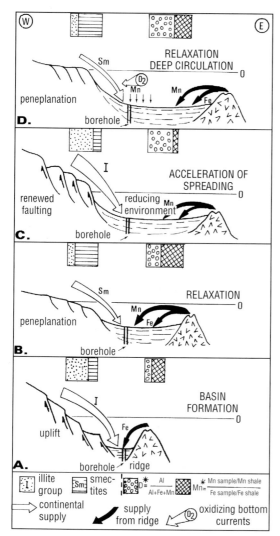

Fig. 7.24 A–D. Oceanic and continental influences on the early evolution of the northwestern Atlantic during the late Jurassic to early Cretaceous, after mineralogical and geochemical data of Debrabant and Chamley (1982). **A** Birth of basin: proximal volcanism (Fe); tectonic uplift of margins and direct erosion of continental rocks (illite group comprising illite, chlorite, random mixed-layers, quartz, various heavy minerals); **B** distal volcanism (Mn); peneplanation of bordering continents and development of hot-climate soils (smectites); **C** temporary rejuvenation of margins (illite); waning of volcanic influences (terrigenous Al, see D* index); reducing marine environment (little Mn precipitated); **D** tectonic quiescence on continents (smectitic soils), development of deep oceanic circulation (precipitation of metal oxides; see Mn* index)

tion of the detrital supply, weakening of diagenetic phenomena, and establishment of exchange between the various bodies of water.

Models taking into account geochemical transfers on a global scale between various types of reservoirs (rocks, soils, oceans, atmosphere) suggest that the evolution of the world climate with time was to a considerable part controlled by the effects of global tectonics (e.g., BLAG-model of Berner et al. 1983; cf. Tardy 1986). In particular, periods of accelerated spreading and thus also of subduction, such as the Cretaceous, correspond to an increased liberation of carbon in the oceans as a result of decarbonation resulting from metamorphism. This carbon became dispersed as CO_2 in the oceanic water and then eventually in the atmosphere where it was responsible for a global warming due to an increased greenhouse effect. The opposite was observed during periods of global cooling, such as after the Eocene, when one had a slow-down of the spreading rates and a decrease of the CO_2 concentration in the marine and atmospheric reservoirs (e.g., Barron 1985). The refinement of these global models, their incorporation into geological and geodynamic concepts, and the consideration of the entire network of effects and causes, should permit an improved understanding of the relations linking the history of the earth's interior with that of its sedimentary cover.

8 References

Académie des Sciences (1984) Les nodules polymétalliques. Faut-il exploiter les mines océaniques? Gauthier-Villars, Paris, 177 pp
Aigner T (1985) Storm depositional systems. Lect Notes Earth Sci 3:174 pp
Allen GP, Laurier D, Thouvenin J (1979) Étude sédimentologique du delta de la Mahakam. Notes Mém CFP 15:156 pp
Allen GP, Castaing P, Tastet JP (1981) Excursion géologique sur l'estuaire de la Gironde. Assoc Sédiment Fr CFP IGBA, 112 pp
Allen JRL (1977) Physical processes of sedimentation. Allen & Unwin, London, 248 pp
Allen JRL (1984) Sedimentary structures. Dev Sedimentol, Elsevier, Amsterdam 30:593, 663
Aloisi JC, Monaco A, Thommeret J, Thommeret Y (1975) Évolution paléogéographique du plateau continental languedocien dans le cadre du Golfe du Lion. Rev Géogr Phys Géol Dyn 2, 17:13–22
Anderson JR, Clark HC, Weaver FM (1977) Sediments and sediment processes on high latitude continental shelves. In: Proc Offshore technology Conf, Houston, pp 91–94
Anderton R (1976) Tidal shelf sedimentation: an example from the Scottish Dalradian. Sedimentology 23:429–458
Arnoux A, Chamley H (1974) Minéraux des argiles et détergents des eaux interstitielles, dans les sédiments superficiels du golfe du Lion. CR Acad Sci Paris 278:999–1002
Badaut D (1981) Néoformation de smectites à partir de frustules de diatomées. Le cas des lacs salés de l'Altiplano de Bolivie. Thèse 3e Cycle, Univ Strasbourg, 73 pp, Ann
Bain DC, Tait JM (1977) Mineralogy and origin of dust fall on skye. Clay Min 12:353–355
Baker PA, Burns SJ (1985) Occurrence and formation of dolomite in organic-rich continental margin sediments. Am Assoc Petrol Geol Bull 69:1917–1930
Bally AW (ed) (1983) Seismic expression of structural styles. Am Assoc Petrol Geol 2 and 3
Barron ES (1985) Explanations of the Tertiary global cooling trend. Palaeo Geogr Climatol Ecol 50:45–61
Baturin GN (1982) Phosphorites on the sea floor. Origin, composition and distribution. Dev Sedimentol, Elsevier, Amsterdam 33:343 pp
Beaudoin B, Fries G, Joseph P, Paternoster B (1983) Sills sédimentaires gréseux injectés dans l'Aptien supérieur de Rosans (Drôme). CR Acad Sci Paris II, 296:387–392
Bellaiche G, Droz L, Aloisi JC, Bouye C, Got H, Monaco A, Maldonado A, Serra-Raventos S, Mirabile L (1981) The Ebro and Rhone deep-sea fans: first comparative study. Mar Geol 43:M75–M85
Berger W (1974) Deep-sea sedimentation. In: Burk CA, Drake CL (eds) The geology of continental margins. Springer, Berlin Heidelberg New York, pp 213–241
Berger W, Winterer JL (1974) Plate stratigraphy and the fluctuating carbonate line. In: Hsü KJ, Jenkyns H (eds) Pelagic sediments on land and under the sea. Spec Publ Int Assoc Sedimentol 1:11–48
Berné S, Augustin JM, Braud F, Chene G, Walker P (1986) Cartographie et interprétation de la dynamique sédimentaire des plates-formes continentales: améliorations de la technique d'observation par sonar latéral. Bull Soc Géol Fr (8) 2:437–446
Berner RA (1971) Principles of chemical sedimentology. McGraw Hill, New York, 240 pp

Berner RA, Lasaga AC, Garrels RM (1983) The carbonate-silicate geochemical cycle and its effect on atmospheric carbon dioxyde over the past 100 million years. Am J Sci 283:641–683

Berners HP (1985) Der Einfluß der Siercker Schwelle auf die Fazies-Verteilungen mesokänozoischer Sedimente im NE des Pariser Beckens. PhD Thesis, Aachen, FRG, 321 pp

Besse D, Desprairies A, Jehanno C (1981) Les paragenèses de smectites et de zéolites dans une série pyroclastique d'âge éocène moyen de l'océan Indien (DSDP leg 26, site 253). Bull Minéral 104:56–63

Biscaye PE (1965) Mineralogy and sedimentation of recent deep-sea clay in the Atlantic Ocean and adjacent seas and oceans. Geol Soc Am Bull 76:803–832

Biscaye PE, Eittreim SL (1977) Suspended particulate loads and transports in the nepheloid layer of the abyssal Atlantic Ocean. Mar Geol 23:155–172

Biscaye PE, Kolla V, Turekian KL (1976) Distribution of calcium carbonate in surface sediments of the Atlantic Ocean. J Geophys Res 81:2595–2603

Blanc JJ (1982) Sédimentation des marges continentales actuelles et anciennes. Masson, Paris, 159 pp

Blatt H (1982) Sedimentary petrology. Freeman, San Francisco, 564 pp

Blatt H, Middleton G, Murray R (1980) Origin of sedimentary rocks. Prentice-Hall, Englewood Cliffs, 782 pp

Boeuf S, Biju-Duval B, De Charpal O, Rognon P, Gariel O, Bennacef A (1971) Les grès du Paléozoïque inférieur au Sahara. Technip, Paris, 464 pp

Boichard R, Burollet PF, Lambert B, Villain J-M (1985) La plate-forme carbonatée du Pater Noster, Est de Kalimantan (Indonésie). Notes Mém CPF 20:103 pp

Boillot G (1979) Géologie des marges continentales. Masson, Paris, 139 pp

Bonatti E (1975) Metallogenesis at oceanic spreading centres. Annu Rev Earth Planet Sci 3:405–431

Bouma AH, Normark WR, Barnes NE (eds) (1985) Submarine fans and related turbidite systems. Springer, Berlin Heidelberg New York, 351 pp

Bouquillon A, Chamley H, Debrabant P, Piqué A (1984) Étude minéralogique et géochimique des forages de Jeumont et Epinoy (Paléozoïque du Nord de la France. Ann Soc Géol N 104:167–179

Bourgeois J (1980) A transgressive shell sequence exhibiting hummocky stratification: the Cape Sebastian sandstone (Upper Cretaceous), south-western Oregon. J Sediment Petrol 50:681–702

Brenner RC (1978) Sussex sandstone of Wyoming, example of Cretaceous offshore sedimentation. Am Assoc Petrol Geol Bull 62:181–200

Brindley GW, Brown G (1980) Crystal structures of clay minerals and their X-ray identification. Mineral Soc, London, 495 pp

Broecker WS, Takahashi T (1966) Calcium carbonate precipitation on the Bahama Banks. J Geophys Res 71:1575–1602

Broussard ML (ed) (1975) Deltas: models for exploration. Houston Geol Soc, Houston, Tex, 555 pp

Bruce CH (1984) Smectite dehydration, its relation to structural development and hydrocarbon accumulation in northern Gulf of Mexico Basin. Am Assoc Petrol Geol Bull 68:673–683

Burk CA, Drake CL (eds) (1974) The geology of continental margins. Springer, Berlin Heidelberg New York, 1009 pp

Burollet PF, Winnock E (ed) (1979) La mer pélagienne. Géol méditerr 6 (1):345 pp

Cadet JP, Kobayashi K et al. (1985) De la fosse du Japon à la fosse des Kouriles: premiers résultats de la campagne océanographique franco-japonaise Kaïko (leg III). CR Acad Sci Paris 301, II:287–296

Caillère S, Hénin S, Rautureau M (1982) Minéralogie des argiles. Masson, Paris, 2 vol 184 pp; and 189 pp

Cailleux A, Tricart J (1959) Initiation à l'étude des sables et des galets, 3 vols. Centre Doc Univ Paris

Caline B (1981) Le secteur occidental de la baie du Mont-Saint-Michel. Doc BRGM Orléans, 42–250 pp

Calvert SE (1974) Deposition and diagenesis of silica in marine sediments. In: Hsü KJ, Jenkins H (eds) Pelagic sediments on land and under the sea. Spec Publ Int Assoc Sedimentol 1:273–299

Carmouze JP, Pédro G, Berrier J (1977) Sur la nature des smectites de néoformation du lac Tchad et leur distribution spatiale en fonction des conditions hydrogéochimiques. CR Acad Sci Paris D 284:615–618

Castaing P (1981) Le transfert à l'océan des suspensions estuariennes. Cas de la Gironde. Thesis Sci Nat Bordeaux I, 530 pp, Ann

Caston VND (1977) Quaternary deposits of the central North Sea. Rep Inst Geol Sci 77 (11):1–8

Chamberlain CK (1975) Trace fossils in DSDP cores of the Pacific. J Paleontol 49:1074–1096

Chamley H (1971) Recherches sur la sédimentation argileuse en Méditerranée. Sci Géol Strasbourg Mém 35:225 pp

Chamley H (1974) Considérations sur la sédimentologie des estuaires. Houille Blanche 1/2:123–128

Chamley H (1981) Long-term trends in clay deposition in ocean. Oceanologica Acta Spec No, pp 105–110

Chamley H, Debrabant P (1984) Paleoenvironmental history of the North Atlantic region from mineralogical and geochemical data. Sediment Geol 40:151–167

Chamley H (1986) Continental and marine paleoenvironments expressed by the West Pacific clay sedimentation. Geol Rundsch 75, 1:271–285

Chamley H (1989) Clay sedimentology. Springer, Berlin Heidelberg New York, 623 pp

Chamley H, Deynoux M, Robert C, Simon B (1980) La sédimentation argileuse du Précambrien terminal au Dévonien dans la région du Hodh (bassin cratonique de Taoudeni, Sud-Est Mauritanien). Ann Soc Géol N 100:73–83

Chamley H, Debrabant P (1984) Paleoenvironmental history of the North Atlantic region from mineralogical and geochemical data. Sediment Geol 40:151–167

Chamley H, Coulon H, Debrabant P, Holtzapffel T (1985) Cretaceous interactions between volcanism and sedimentation in the Mariana Basin, from mineralogical, micromorphological and geochemical investigations (site 585 DSDP). Init Rep DSDP 89:413–429

Cline RM, Hays JD (eds) (1976) Investigation of late Quaternary paleoceanography and paleoclimatology. Geol Soc Am Mem 145:105 pp

Cole TG (1985) Composition, oxygen isotope geochemistry, and origin of smectite in the metalliferous sediments of the Bauer Deep, South-East Pacific. Geochim Cosmochim Acta 42:221–235

Coleman JM (1981) Deltas: processes of deposition and models for exploration. Burgess, Champaign, Ill, 124 pp

Collinson JD, Thompson DB (1982) Sedimentary structures. Allen & Unwin, London Boston, 194 pp

Cooke RU, Warren A (1973) Geomorphology in deserts. Univ Cal Press, Berkeley, 374 pp

Coudé-Gaussen G (1982) Les poussières éoliennes sahariennes. Mise au point. Rev Géomorphol Dyn 31:49–69

Cremer M (1985) La levée nord de l'éventail subaquatique profond du Cap Ferret: rôle du modèle du fond et de l'accélération complémentaire de Coriolis sur la répartition des dépôts de débordements turbiditiques. Bull Soc geol Fr (8) 1:49–57

Dallery H (1955) Les rivages de la Somme. Mém Soc Em Hist Litt, Abbeville 14:308 pp

Davies JL (1973) Geographical variation in coastal development. Hafner, New York, 204 pp

Davis RA (ed) (1978) Coastal sedimentary environments. Springer, Berlin Heidelberg New York, 420 pp

Dean WE, Leinen M, Stow DAV (1985) Classification of deep-sea, fine-grained sediments. J Sediment Petrol 55:250–256

Debrabant P, Chamley H (1982) Influences océaniques et continentales dans les premiers dépôts de l'Atlantique Nord. Bull Soc Géol Fr (7) 24:473–486

Deconinck J-F, Debrabant P (1985) Principaux aspects de la diagenèse des argiles dans le domaine subalpin. Rev Géol Dyn Géogr Phys 26:321–330

Despeyroux Y (1985) Etude hydrosédimentaire de l'estuaire de la Canche (Pas-de-Calais). Thesis 3ᵉ Cycle, Lille I, 188 pp, Ann

Desprairies A, Jehanno C (1983) Paragenèses minérales liées à des interactions basalte-sédiment-eau de mer (sites 465 et 456 des legs 65 et 60 du DSDP). Sci Géol Bull Strasbourg 36:93–110

Deynoux M (1978) Les formations glaciaires du Précambrien terminal et de la fin de l'Ordovicien en Afrique de l'Ouest. Deux exemples de glaciation d'inlandsis sur une plate-forme stable. Thesis Sci Nat Aix-Marseille III, 554 pp

Deynoux M (1985) Terrestrial or waterlain glacial diamictites? Three case studies from the late Precambrian and late Ordovician glacial drifts in West Africa. In: Deynoux M (ed) Glacial record. Palaeogeogr Climatol Ecol Spec Issne 51:97–141

Donovan RN (1975) Devonian lacustrine limestones at the margin of the Arcadian Basin, Scotland. J Geol Soc London 131:489–510

Doyle LJ, Pilkey OH (eds) (1979) Geology of continental slopes. Soc Econ Palaentol Mineral Tulsa Spec Publ 27

Dunham RJ (1962) Classification of carbonate rocks according to depositional texture. Am Assoc Petrol Geol, Mem 1:108–121

Einsele G, Seilacher A (eds) (1982) Cyclic and event stratification. Springer, Berlin Heidelberg New York, 536 pp

Embry AF, Klovan JE (1971) A late Devonian reef tract on northeastern Banks Island. NWT Bull Can Petrol Geol 19:730–781

Eugster HP, Harde LA (1975) Sedimentation in an ancient playa-lake complex: the Wilkins Peak Member of the Green River Formation of Wyoming. Geol Soc Am Bull 86:319–334

Fairbridge RW, Bourgeois J (eds) (1978) The encyclopedia of sedimentology. Dowden, Hutchinson & Ross, Stroudsburg, 901 pp

Faugères JC, Stow DAV, Gonthier E (1984) Contourite drift moulded by deep Mediterranean outflow. Geology 12:296–300

Ferm JC (1974) Carboniferous environment models in eastern United States and their significance. In: Briggs G (ed) Carboniferous of the SE United States. Geol Soc Am Spec Pap 148:79–96

Fisher RV, Schmincke HU (1984) Pyroclastic rocks. Springer, Berlin Heidelberg New York, 470 pp

Fisk HN (1944) Geological investigations of the alluvial valley of the Lower Mississippi Valley. Miss River Commiss Vicksburg, 78 pp

Flemming BW (1980) Sand transport and bedform patterns on the continental shelf between Durban and Port Elizabeth (southeast Africa continental margin). Sediment Geol 26:179–205

Flint RF (1971) Glacial and quaternary geology. John Wiley & Sons, New York, 892 pp

Folk RL (1966) A review of grain-size parameters. Sedimentology 6:73–93

Folk RL (1980) Petrology of sedimentary rocks. Hemphills, Austin, 182 pp

Folk RL, Land LS (1975) Mg:Ca ratio and salinity: two controls over crystallization of dolomite. Am Assoc Petrol Geol Bull 59:60–68

Frakes LA (1979) Climate throughout geologic time. Elsevier, Amsterdam, 310 pp

Frey RW (ed) (1975) The study of trace fossils. Springer, Berlin Heidelberg New York, 562 pp

Friedman GM (1961) Distinction between dune, beach and river sands from their textural characteristics. J Sediment Petrol 31:514–529

Friedman GM, Sanders JE (1978) Principles of sedimentology. John Wiley & Sons, New York, 792 pp

Fries G, Beaudoin B (1984) L'éventail sous-marin de Ceüse au Gargasien (Drôme, Hautes-Alpes). 10ème Réun Ann Sci Terre, Bordeaux:233

Fryberger SG, Ahlbrandt TS, Andrews S (1979) Origin, sedimentary features and significance of low-angle eolian "sand-sheet" deposits. Great sand dunes, National Monument and vicinity, Colorado. J Sediment Petrol 49:733–746

Gall JC (1976) Environnements sédimentaires anciens et milieux de vie. Doin, Paris, 233 pp

Galloway WE, Hobday DK (1983) Terrigenous clastic depositional systems. Springer, Berlin Heidelberg New York, 423 pp

Garrels RM, Mackenzie FT (1971) Evolution of sedimentary rocks. Norton, New York, 397 pp

Gebelein CD (1969) Distribution, morphology and accretion rate of recent subtidal algal stromatolites, Bermuda. J Sediment Petrol 39:9–69

Gibbs RJ (ed) (1974) Suspended solids in water. Marine science, vol 4. Plenum, New York, 320 pp

Gibbs RJ (1977) Clay mineral segregation in the marine environment. J Sediment Petrol 47:237–243

Ginsburg RN (1964) South Florida carbonate sediments. Geol Soc Am Annu Convent, Field guidebook: 71 pp

Glasby GP (ed) (1977) Marine manganese deposits. Elsevier, Amsterdam, 523 pp

Glennie KW (1970) Desert sedimentary environments. Elsevier, Amsterdam, 222 pp

Glennie KW (1972) Permian Rotliegendes of northwest Europe interpreted in the light of modern desert sedimentation studies. Bull Am Assoc Petrol Geol 56:1048–1071

Goldthwait RP (ed) (1975) Glacial deposits. Dowden, Hutchinson & Ross, Stroudsburg: 464 pp

Gould HR, McFarlan E (1959) Geological history of the chenier plain, southwestern Louisiana. Trans Gulf-Coast Assoc Geol Soc 9:261–270

Graciansky PC, Brosse E, Deroo G, Herbin JP, Montadert L, Müller C, Sigal J, Schaaf A (1982) Les formations d'âge crétacé de l'Atlantique Nord et leur matière organique: paléogéographie et milieux de dépôt. Rev Inst Fr Pétrole 37:275–337

Griffin JJ, Windom H, Goldberg ED (1968) The distribution of clay minerals in the world ocean. Deep Sea Res 15:433–459

Guilcher A (1979) Précis d'hydrologie marine et continentale. Masson, Paris, 344 pp

Hallam A, Sellwood BW (1976) Middle Mesozoic sedimentation in relation to tectonics in the British area. J Geol 84:301–321

Hardie LA (ed) (1977) Sedimentation on the modern carbonate tidal flats of NW Andros Island, Bahamas. Univ Press, John Hopkins, Baltimore

Hardie LA, Smoot JP, Eugster HP (1978) Saline lakes and their deposits: a sedimentological approach. In: Matter A, Tucker ME (eds) Modern and ancient lake sediments. Spec Publ Int Assoc Sedimentol 2:7–42

Hays JD, Imbrie J, Shackleton NJ (1976) Variations in the Earth's orbit: pacemaker of the ice ages. Science 194:1121–1132

Heath GR (1974) Dissolved silica and deep-sea sediments. In: Gay NW (ed) Studies in paleoceanography. Soc Econ Palaeontol Mineral (Spec Publ) 20:77–93

Heezen BC, Hollister CD (1971) The face of the deep. Univ Press, Oxford, 659 pp

Heezen BC, Hollister CD, Ruddiman WF (1966) Shaping of the continental rise by deep geostrophic contour currents. Science 152:502–508

Heward AP (1978) Alluvial fan sequence and megasequence models, with examples from Westphalian D-Stephanian B coalfields, northern Spain. In: Miall AD (ed) Fluvial sedimentology. Mem Can Soc Petrol Geol 5:669–702

Hoffert M (1980) Les "argiles rouges des grands fonds" dans le Pacifique centre-est. Authigenèse, transport, diagenèse. Sci Géol Strasbourg Mém 61:231 pp

Hoffert M (1985) Les fonds océaniques: nouvelles frontières de la recherche minière? Bull Soc Géol Fr (8) 1:979–990

Holtzapffel T, Bonnot-Courtois C, Chamley H, Clauer N (1985) Héritage et diagenèse des smectites du domaine sédimentaire nord-atlantique (Crétacé, Paléogène). Bull Soc Géol Fr (8) 1:25–33

Honjo S, Manganini SJ, Poppe LJ (1982) Sedimentation of lithogenic particles in the deep ocean. Mar Geol 50:199–220

Horowitz AS, Potter PE (1971) Introductory petrography of fossils. Springer, Berlin Heidelberg New York, 302 pp

Houbolt JJHC, Jonker JBM (1968) Recent sediments in the eastern part of the Lake of Geneva (Lac Léman). Geol Mijnbouw 47:131–148

Hsü KJ, Jenkyns HC (eds) (1974) Pelagic sediments: on land and under the sea. Spec Publ Int Assoc Sedimentol: 447 pp

Hsü JJ, Montadert L et al. (1978) Initial Reports of the Deep Sea Drilling Project 42A:1249 pp

Jouanneau JM (1982) Matières en suspension et oligo-éléments métalliques dans le système estuarien girondin: comportement et flux. Thesis Sci Nat Bordeaux I:150 pp, dact, Ann

Keller J, Ninkovich D (1972) Tephra-Lagen in der Ägäis. Z Dtsch Geol Ges 123:579–587

Kelts K, Hsü HJ (1978) Freshwater carbonate sedimentation. In: Lerman A (ed) Lakes: physics, chemistry and geology. Springer, Berlin Heidelberg New York, pp 295–321

Kennett JP (1982) Marine geology. Prentice Hall, Englewood Cliffs, 813 pp

Kirkland DW, Evans R (1973) Marine evaporites: origins, diagenesis and geochemistry. Dowden, Hutchinson & Ross, Stroudsburg, 426 pp

Komar PD (1976) Beach processes and sedimentation. Prentice Hall, Englewood Cliffs

Komar PD (1983) Handbook of coastal processes and erosion. CRC Press, Boca Raton, 305 pp

Krauskopf KB (1967) Introduction to geochemistry. McGraw Hill, New York, 706 pp

Lancelot Y (1978) Relations entre évolution sédimentaire et tectonique de la plaque pacifique depuis le Crétacé inférieur. Mém Soc Géol Fr 134:164 pp

Larsen G, Chilingar GV (eds) (1979) (1983) Diagenesis in sediments and sedimentary rocks. Elsevier, Amsterdam, vols 1, 2. 579 pp, and 572 pp

Larsonneur C (1975) Normandie, baie du mont Saint-Michel, France. In: 9th Int Congr Sédimentologie, Nice, 128 pp

Larsonneur C, Bouysse P, Auffret JP (1982) The superficial sediments of the English Channel and its western approaches. Sedimentology 29:851–864

Lauff GH (ed (1967) Estuaries. Am Assoc Adv Sci Washington 83:757 pp

Leatherman SP (ed) (1979) Barrier islands. Academic Press, New York London, 325 pp

Leeder MR (1982) Sedimentology. Process and product. Allen & Unwin, London Boston, 344 pp

Lees A (1975) Possible influences of salinity and temperature on modern shelf carbonate sedimentation. Mar Geol 19:159–198

Lenôtre N, Chamley H, Hoffert M (1985) Nature and significance of sites 576 and 578 clay stratigraphy (leg 86 DSDP, north-western Pacific). Init Rep DSDP 86:571–579

Le Pichon X, Ilyama T et al. (1985) La subduction du bassin de Shikoku et de ses marges le long du fossé de Nankaï (Japon méridional): résultats préliminaires du programme Kaïko (leg 1). CR Acad Sci Paris II, 301:273–280

Le Ribault L (1977) L'exoscopie des quartz. Masson, Paris, 150 pp

Lerman A (ed) (1978) Lakes: physics, chemistry and geology. Springer, Berlin Heidelberg New York, 366 pp

Lisitzin AP (1972) Sedimentation in the world ocean. Soc Econ Palaeontol Mineral Tulsa, Spec Publ 17:218 pp

Loreau JP (1982) Sédiments aragonitiques et leur genèse. Mém Mus Nat Hist Nat Paris, 47:312 pp

Lucas J (ed) (1979) Phosphorites. Sci Géol Strasbourg 32:3–105

Lucas J, Prévôt L (1975) Les marges continentales, pièges géochimiques; l'exemple de la marge continentale de l'Afrique à la limite Crétacé-Tertiaire. Bull Soc Géol Fr (7) 17:496–501

Manickam S, Barbaroux L, Ottmann F (1985) Composition and mineralogy of suspended sediments in the fluvio-estuarine zone of the Loire River, France. Sedimentology 32:721–741

Marchig V (1981) Marine manganese nodules. Top Curr Chem (Fortschr Chem Forsch) 25:99–126

Matter A, Tucker MA (ed) (1978) Modern and ancient lake sediments. Spec Publ Int Assoc Sedimentol 2:290 pp

McCoy FW, Zimmerman HB (1977) A history of sediment facies in the south Atlantic Ocean. Init Rep DSDP 39:1047–1079

McGowen JH, Groat CG (1971) Van Horn sandstone, West Texas: an alluvial fan model for mineral exploration. Bur Econ Geol Univ Texas, Austin, Rep Invest 72:57 pp

Melguen M, Le Pichon X, Sibuet J-C (1978) Paléoenvironnement de l'Atlantique Sud. Bull Soc Géol Fr (7) 20:471–490

Mélières F (1973) Les minéraux argileux de l'estuaire du Guadalquivir (Espagne). Bull Gr Fr Argiles 25:161–172

Miall AD (ed) (1978) Fluvial sedimentology. Mem Can Soc Petrol Geol 5:859 pp

Miall AD (1984) Principles of sedimentary analysis. Springer, Berlin Heidelberg New York, 490 pp

Milliman JD (1974) Marine carbonates. Springer, Berlin Heidelberg New York, 375 pp

Millot G (1970) Geology of clays. Springer, Berlin Heidelberg New York, 425 pp

Millot G (1980) Les grands aplanissements des socles continentaux dans les pays subtropicaux, tropicaux et désertiques. Mém Soc Géol Fr Spéc Publ 10:295–305

Moore DG (1969) Reflection profiling studies of the California continental borderland: structure and quaternary turbidite basins. Geol Soc Am Spec Pap 107:1–142

Moore GT, Starke GW, Bonham LC, Woodbury HO (1978) Mississippi Fan, Gulf of Mexico, physiography, stratigraphy and sedimentational patterns. In: Moore GT, Coleman JM (eds) Studies in geology. Soc Econ Paleontol Mineral, Tulsa, 7:155–191

Morton RA (1972) Clay mineralogy of Holocene and Pleistocene sediments, Guadalupe delta of Texas. J Sediment Petrol 42:85–88

Munk WH, Traylor MA (1947) Refraction of ocean waves, a process linking underwater topography to beach erosion. J Geol 55:1

Nesteroff WD, Mélières F (1967) L'érosion littorale du pays de Caux. Bull Soc Géol Fr (7) 9:159–169

Odin GS, Matter A (1981) De glauconiarum origine. Sedimentology 28:611–641

Oomkens E (1970) Depositional sequences and sand distribution in the post-glacial Rhone delta complex. In: Deltaic sedimentation: modern and ancient. Soc Econ Palaeontol Mineral Tulsa Spec Publ 15:198–212

Park RK (1977) The preservation potential of some recent stromatolites. Sedimentology 24:485–506

Pautot G, Hoffert M (1984) Les nodules du Pacifique central dans leur environement géologique. Publ CNEXO Rés Camp Mer 26:202 pp

Pédro G (1968) Distribution des principaux types d'altération chimique à la surface du globe. Présentation d'une esquisse géographique. Rev Géol Dyn Géogr Phys 10,5:457–470

Pédro G (1979) Caractérisation générale des processus de l'altération hydrolytique. Science du Sol. Bull AFES 2–3:93–105

Pelet R (1985) Sédimentation et évolution géologique de la matière organique. Bull Soc Géol Fr (8) 1:1075–1086

Pérès JM (1961) Océanographie biologique et biologie marine. I. La vie benthique. Presses Univ Fr, Coll Euclide, Paris, 541 pp

Perrodon A (1985) Géodynamique pétrolière. Masson & Elf, Paris, 385 pp

Pettijohn FJ (1975) Sedimentary rocks. Harper & Row, New York, 628 pp

Pettijohn FJ, Potter PE (1964) Atlas and glossary of sedimentary structures. Springer, Berlin Heidelberg Göttingen, 370 pp

Picard J (1967) Essai de classement des grands types de peuplements benthiques tropicaux, d'après les observations effectuées dans les parages de Tuléar (SW de Madagascar). Rec Trav Stn Mar End Spec Publ 6:3–24

Picard MD, High LR Jr (1981) Physical stratigraphy of ancient lacustrine deposits. Soc Econ Palaeontol Mineral Tulsa Spec Publ 31:233–259

Pilkey OH, Locker SD, Cleary WJ (1980) Comparison of sand-layer geometry on flat floors of 10 modern depositional basins. Am Assoc Petrol Geol Bull 64:841–856

Potter PE, Pettijohn FJ (1977) Paleocurrents and basin analysis. Springer, Berlin Heidelberg New York, 425 pp

Potter PE, Maynard JB, Pryor WA (1980) Sedimentology of shale. Springer, Berlin Heidelberg New York, 306 pp

Purser BH (1980) Sédimentation et diagenèse des carbonates néritiques récents, vol 1. Technip, Paris, 366 pp
Purser BH (1983) Sédimentation et diagenèse des carbonates néritiques récents, vol 2. Technip, Paris, 389 pp
Reading H (ed) (1986) Sedimentary environments and facies. Blackwell, Oxford, 636 pp
Reineck HE, Singh IB (1980) Depositional sedimentary environments. Springer, Berlin Heidelberg New York, 549 pp
Renard H (1986) Pelagic carbonate chemostratigraphy (Sr, Mg, 18_O, 13_C). Mar Micropalaeontol 10:117–164
Richter-Bernberg G (ed) (1972) Geology of saline deposits. UNESCO, Paris
Riech V, von Rad U (1979) Silica diagenesis in the Atlantic ocean: diagenetic potential and transformations. In: M Ewing Series, vol 3. Am Geophys Un:315–340
Rivière A (1977) Méthodes granulométriques. Techniques et interprétations. Masson, Paris, 170 pp
Robert C (1982) Modalité de la sédimentation argileuse en relation avec l'histoire géologique de l'Atlantique Sud. Thesis Sci Nat Aix-Marseille II:141 pp, Ann
Roberts DG, Kidd RB (1979) Abyssal sediment wave fields on Feri Ridge, Rockall Trough: long-range sonar studies. Mar Geol 33:175–191
Robertson AHF, Bliefnick DM (1983) Sedimentology and origin of lower Cretaceous pelagic carbonates and redeposited clastics, Blake-Bahama formation, Deep Sea Drilling Project, site 534, Western Equatorial Atlantic. Init Rep DSDP 76:795–828
Rona PA (1984) Hydrothermal mineralization at seafloor spreading centers. Earth Sci Rev 20:1–104
Ruddiman WF (1977) Late quaternary deposition of ice-rafted sand in the subpolar North Atlantic (lat. 40° to 65° N). Geol Soc Am Bull 88:1813–1827
Salomon JC, Allen GP (1983) Rôle de la marée dans les estuaires à fort marnage. Notes Mém CFP 18:35–44
Sarnthein M, Seibold E, Rognon P (eds) (1980) Sahara and surrounding seas. Palaeoecology of Africa, vol 12. Balkema, Rotterdam, 408 pp
Schlanger SO, Cita MB (eds) (1982) Nature and origin of Cretaceous carbon-rich facies. Academic Press, London New York, 229 pp
Scholle PA, Schulger PR (eds) (1979) Aspects of diagenesis. Soc Econ Paleontol Mineral, Tulsa, Spec Publ 26:400 pp
Scholle PA, Spearing D (eds) (1982) Sandstone depositional environments. Am Assoc Petrol Geol Mem 31:401 pp
Scholle PA, Bebout DG, Mocre CH (1983) Carbonate depositional environments. Am Assoc Petrol Geol Mem 33:708 pp
Schumm SA (1977) The fluvial system. John Wiley & Sons, New York, 338 pp
Seibold E, Berger WH (1982) The sea floor. An introduction to marine geology. Springer, Berlin Heidelberg New York, 288 pp
Selley RC (1976) An introduction to sedimentology. Academic Press, New York London, 408 pp
Selley RC (1978) Ancient sedimentary environments. Chapman & Hall, London, 237 pp
Sims JD (1978) Annoted bibliography of penecontemporaneous deformational structures in sediments. US Geol Surv Open File Rep 78 (510):1–78
Singer A (1984) The paleoclimatic interpretation of minerals in sediments. A review. Earth Sci Rev 21:251–293
Singer A, Stoffers P (1980) Clay mineral diagenesis in two east African lake sediments. Clay Minerals 15:291–307
Slansky M (1980) Géologie des phosphates sédimentaires. Mém BRGM 114:92 pp
Steel RJ, Aasheim SM (1978) Alluvial sand deposition in a rapidly subsiding basin (Devonian, Norway). In: Miall AD (ed) Fluvial sedimentology. Mem Can Soc Petrol Geol 5:385–412
Steinberg M, Touray JC, Treuil M, Massard P (1978) (1979) Géochimie. Principes et méthodes. Doin, Paris, vols 1, 2. 599 pp

Sternberg RW, Larsen LH (1976) Frequency of sediment movement on the Washington continental shelf: a note. Mar Geol 21:M37–M47
Stow DAV, Lovell JPB (1979) Contourites: their recognition in modern and ancient sediments. Earth Sci Rev 14:251–291
Stow DAV, Piper DJW (eds) (1984) Fine-grained sediments: deep-water processes and facies. Geol Soc. Blackwell, London, 659 pp
Stow DAV, Howell DG, Nelson CH (1983/1984) Sedimentary, tectonic and sea-level controls on submarine fans and slope-apron turbidite systems. Geo Mar Lett 3:57–64
Sturm M, Matter A (1978) Turbidites and varves in Lake Brienz (Switzerland): deposition of clastic detritus by density currents. In: Matter A, Tucker ME (eds) Modern and ancient lake sediments. Spec Publ Int Assoc Sedimentol 2:145–166
Surdam RC, Eugster AP (1976) Mineral reactions in the sedimentary deposits of the Lake Magadi region, Kenya. Geol Soc Am Bull 87:1739–1752
Taira A, Niitsuma N (1985) Turbidite sedimentation in Nankai Trough as interpreted from magnetic fabric, grain size and detrital mode analysis. Init Rep DSDP 87 A:611–632
Tardy Y (1986) Le cycle de l'eau. Masson, Paris, 338 pp
Thiede J, van Andel TH (1977) The paleoenvironment of anaerobic sediments in the late Mesozoic South Atlantic Ocean. Earth Planet Sci Lett 33:301–309
Tiercelin JJ, Périnet G, Le Fournier J, Briéda S, Robert P (1982) Lacs du rift est-africain, exemples de transition eaux douces-eaux salées: le lac Bogoria, rift Gregory, Kenya. Mém Soc Géol Fr 41, 144:217–230
Tissot BP, Welte DH (1978) Petroleum formation and occurrence. Springer, Berlin Heidelberg New York, 538 pp
Tucker ME (1982) Sedimentary petrology. An introduction. Blackwell, Oxford, 252 pp
Vanney JR (1977) Géomorphologie des plates-formes continentales. Doin, Paris, 300 pp
Vatan A (1967) Manuel de sédimentologie. Technip, Paris, 397 pp
Verger F (1983) Marais et Wadden du littoral français. Minard, Paris, 549 pp
Walker RG (1978) Deep water sandstone facies and ancient submarine fans: models for exploration for stratigraphic traps. Am Assoc Petrol Geol Bull 62:932–966
Walker RG (ed) (1984) Facies models. Geosci Can, Reprint Series 1:317 pp
Warme JE, Douglas RG, Winterer EL (1981) The Deep Sea Drilling Project: a decade of progress. Soc Econ Palaeontol Mineral Tulsa Spec Publ 32:564 pp
West IM (1965) Evaporites and associated sediments of the basal Purbeck formation (Upper Jurassic) of Dorset. Proc Geol Assoc 86:205–225
Weydert P (1973) Les formations récifales de la région de Tuléar (côte SW de Madagascar). Aperçu de leurs aspects morphologiques, sédimentologiques et de leur évolution. Assoc Sénég Et Quat Ouest Afr 37–38:57–82
Wiley M (ed) (1976) Estuarine processes, 2 vols. John Wiley & Sons, New York Chichester
Wilson IG (1971) Desert sand flow basins and a model for the development of ergs. Geogr J 137:180–199
Wilson JC (1975) Carbonate facies in geologic history. Springer, Berlin Heidelberg New York, 471 pp
Wollast J (1974) The silica problem. In: Goldberg ED (ed) The sea. John Wiley & Sons, New York, p 5
Wright WB (1937) The quaternary Ice Age. Macmillan, London, 478 pp
Zenger DH, Dunham JB, Ethington RL (eds) (1980) Concepts and models of dolomitization. Soc Econ Palaeontol Mineral Tulsa Spec Publ 28:320 pp
Ziegler PA (1975) North Sea Basin history in the tectonic framework of NW Europe. In: Woodland AW (ed) Petroleum and the continental shelf of NW Europe. Appl Sci Publ, vol 1, Geology, pp 131–149

Subject Index

A
abyssal plain   229, 240
accretionary prism   242
acidolysis   3
adhesion ripple   78
aeolian flux   139, 261
aggradation   157, 178, 195, 214, 239
aggregate   13, 22, 62, 140, 171, 251
alcalinolysis   3
algae   19, 148
algal mat   20, 198, 201
alitization   6, 9
allochem   13
alluvial plain   158
analcite   112
anchizone   107, 127
anhydrite   41, 198
anthracite   125
antidune   67
antiripplet   78
apatite   38
aquatill   137
aragonite   14, 16, 21, 118, 220
arenite   48

B
back-shore   189
bacterial activity   20, 38, 106, 127, 198, 255
bafflestone   222
bahamite   21
ball-and-pillow   82
bank   159, 193, 211
bar   157, 160, 173, 190, 193
barchan   141
barrier island   186, 191
beach rock   110, 118
bedding   67, 73
berm   190
bindstone   222
bioclast   13, 16, 219
biocoenose   189
bioherm   222

biosiliceous   247, 264
biostrome   222
bioturbation   83, 97, 208
bird-eye   79, 118, 221
bischofite   43
bisialitization   6, 8
bituminous   122
black shale   127
bone bed   38
borate   43, 150
bottom set   167
Bouma sequence   91, 236
boundstone   12

C
calcite   14, 21, 110, 117, 201
canyon   229, 233
carbonate compensation depth (CCD)   25, 253, 264
carbonates   12, 16, 116, 151, 197, 201, 215, 244, 264
carnallite   41
cast   66
cementation   106, 110, 117
chalk   59
channel   79, 157, 161, 172, 233
chenier   196
chert   35, 59, 123
chicken wire   42
chloralgal   24
chlorite   4, 110, 113, 250
chlorozoan   23
chute and pool   72
classification   57, 83, 108
clay, claystone   48, 58, 111, 114, 249
clay minerals   3, 110, 148, 171, 211, 249
clay plug   170
climate   6, 250
climbing ripple   73, 135
clinoptilolite   112
coal   124, 197
coal rank   125

coastal barrier 179
coastal environment 184
coating 109, 142, 262
compaction 107, 175
concretion 106
cone-in-cone 120
contourite 87, 94, 96, 232
convolute bedding 82
crescent 78
crevasse splay 158, 174
cross bedding 73, 140, 222
current 204
current ripple 66
cyclothem 162

D
debris flow 64, 91, 137, 154, 235
dehydration 107, 115
delta 147, 167, 191
desert 139
desiccation 81
diagenesis 34, 105
diamictite 133
diatomite 35
dish 82
dissolution 106, 110
distal 100, 155, 164, 229, 238
dolomite 14, 16, 42, 59, 110, 121, 152
drainage 8
dreikanter 140
drift 232
dropstone 137
drumlin 133
dune 140, 189
dust 140, 260, 261
dyke 80

E
enterolithic 42
epigenesis 107
epilimnion 145
erg 141
erosion 75, 132, 140
esker 135
estuary 167
euxinic 40
evaporites 1, 41, 112, 142, 144, 148, 152, 197
exoscopy 54
extraclast 13

F
fan 153, 163, 198, 233
fenestrae 118
flaser 74
floatstone 12, 222

flocculate 62, 171
flow of particles 64, 67, 89, 91
fluid migration 242, 257
fluid mud 171
flute 75
fluvio-glacial 134
foramol 23
foreset 167
foreshore 190
framestone 222

G
geochemical transfer 257, 269
glacial 134
glacier 131, 262
glacio-lacustrine 134
glacio-marine 134, 262
glauconite 5, 112
graded-bedding 74, 235
grain fabrics 55
grain flow 64, 91
grain shape 54
grain size 47, 49, 61, 164
grainstone 12, 223
grapestone 23
gravity flow 90
graywacke 110
groove 78
guano 38
guyot 230
gypsum 41, 198

H
halite 41, 148
halloysite 5
halmyrolysis 106
halokinesis 45
hamada 141
hardground 221
hemipelagite 86, 100
herring-bone 74
hummocky 74
hydrocarbon 125, 127
hydrolysis 2
hydrothermalism 148, 256
hypolimnion 146

I
ichnofossil 83
illite 4, 110, 113, 249
impact 79
inertinite 124
infiltration-reflux 121
infratidal 190, 199
inlet 191
intertidal 190, 198

Subject Index 283

intra-oceanic basin 230, 241
inundite 96

K
kainite 43
kames 135
kaolinite 4, 7, 110, 249
kerogen 127
knoll 215

L
lag deposit 159
lagoon 195, 198, 220
lake 144
lamination, laminite 20, 40, 45, 67, 73, 89, 93, 134, 138, 146
lenticular bedding 74
levee 173, 233
lignite 125
limestone 12, 59, 119
lineation 79
liptinite 124
liquefaction 82
liquefied flow 64, 91
lithoclast 13
littoral barrier 194, 197
littoral transfer 189, 193
load structure 82
lobe 174, 229, 236
locomorphic 107
lump 23
lussatite 34, 59, 123
lysocline 25

M
maceral 124
macro-, meso-, microtidal 185
margin 227, 231
mark 66
marl 59
marsh 191
matrix 13
megaripple 67
metalliferous 253, 256
methane 127
micrite 13, 19, 117
mixed-layer 4, 250
mixtite 134
mode 49
monosialitization 6
moraine 132
morphoscopy 54
mud, mudstone 12, 48, 91, 108, 114, 211
mud ball 23
mud-cracks 2, 81
mud diapir 175
mud mound 30, 215
mud volcano 79
mud wave 232

N
natural gaz 127
neomorphism 117
nepheloid layer 63, 94
nodule 106, 120, 242, 253

O
offlap 213
offshore 190
oil window 128
olistostrome 91
oncoid 20
oncolite 20, 21
onlap 213
ooid 13, 21
oolite 21, 198, 219
opal 122
organic matter 38, 124, 148, 182
outwash 135, 138
ouze 13, 18, 25, 32, 58, 114, 244, 249, 264
oxygen minimum zone 40

P
packstone 12
paleocurrent 102
paleodepth 101
paleoenvironment 99, 137, 144, 150, 154, 162, 182, 197, 202, 222, 264
paleogeography 9, 30, 138, 214, 265
palygorskite 4, 112, 140, 250
peat 125
pediment 141
pelagite 86, 110, 242
pelite 48
pellet 13, 22, 24, 198, 201, 220, 246, 251
peloid 19, 21
periodite 88, 97
permeability 57, 111, 129
petroleum 127
philippsite 112
phi-scale 49
phosphate 36, 38, 248
phosphorite 35
phyllomorphic 107
pisolite 21
pit-and-mound 79
placer deposit 197
plane bed 68, 71
plate stratigraphy 263
platform 12, 19, 29, 219
playa 142, 148
podzolization 8

polyhalite 41
porcellanite 34, 59, 123
porosity 57, 108, 110, 113
precipitation 15, 37, 42, 172, 255
prodelta 173
productivity 27, 32, 37, 148, 208, 247, 264
progradation 178, 192, 195, 225, 239
proximal 100, 155, 164, 229, 236
pyroclastic 259
pyrophyllite 4

Q
quartz 9, 54, 109, 122, 140, 262

R
radiolarite 35, 59
recrystallization 107
red clay 25, 256, 264
redomorphic 107
reduction 39, 113
reef 196, 215, 216, 219, 222
reg 141
regression 195, 213
resedimentation 88, 175, 232, 242
reservoir 130
rhythmite 88, 134, 148
ridge 230
rill 79
rip current 188
ripple 67, 140
rise 229
river 157
rock flour 132
rolling 62
roundness 54, 140
rudist reef 225
rudite 48
rudstone 12, 222

S
sabkha 42, 142, 197, 202
saline wedge 170
salinolysis 3, 9
saltation 62
sand, sandstone 48, 59, 107, 140
sand ribbon 209
sandur 135
sand wave 67, 209
scour 75, 79
seamount 230
sedimentation rate 249, 255
seiche 145
seif 141
sepiolite 4, 112
serpentine 5
settling 86, 171, 212, 241

shale 114
shelf 204, 227
shoreface 190
silica 30, 109, 122, 247
sill 80
silt, siltstone 48, 59
skeleton 16, 23
skewness 50, 140
slope 205, 229, 234
slump, slumping 64, 82, 88, 175, 235
smectite 4, 110, 112, 129, 148, 171, 249, 250, 257, 260
smoker 256
soil 2, 3
sorting 49, 140, 171, 251, 260
sparite 13, 117
sphericity 54
spherulite 21
stromatolite 20, 198, 201, 221, 226
stylolite 120
subsidence 175, 180, 214, 226, 263
sulfide 257
supratidal 189, 198
suspension 62, 171
swash 79
sylvite 41
synaeresis 81

T
tar 130
tectonic control 242, 263
tempestite 96, 206
tephra 259
terrace 162
terrigenous 1, 10, 147, 209, 233, 242, 250, 267
tidal flat 167, 190, 192, 195, 197, 221
tidal ridge 181, 211
tide 167, 180, 185
till 133
tombolo 193
tool mark 78
topset 167
transgression 179, 194, 213, 239
transport 61, 65
travertine 148
trench 231, 242
trona 41, 148
tufa 148
turbidite 87, 89, 91, 96, 100, 137, 235, 240, 242
turbidity 62, 64, 91, 170

U
upwelling 36, 208, 247

V
varve 137
velocity 61
vermiculite 4
vitrinite 124
volcanogenic 251, 259, 267

W
wacke 108
wackestone 12

washover 194
water stratification 39, 145
wave reflexion, refraction 186, 189
wave ripple 72
weathering 1, 2, 5, 8, 139
whiting 16, 221
wildflysch 91
wrinkle 79

Z
zeolite 112, 148, 251, 258, 261